Lecture Notes in Mathematics

A collection of informal reports and seminars
Edited by A. Dold, Heidelberg and B. Eckmann, Zürich

T0233368

327

Eben Matlis
Northwestern University, Evanston, IL/USA

1-Dimensional Cohen-Macaulay Rings

Springer-Verlag
Berlin · Heidelberg · New York 1973

AMS Subject Classifications (1970): 13-02, 13Cxx, 13E05, 13E10, 13F05, 13Hxx

ISBN 3-540-06327-7 Springer-Verlag Berlin · Heidelberg · New York
ISBN 0-387-06327-7 Springer-Verlag New York · Heidelberg · Berlin

Offsetdruck: Julius Beltz, Hemsbach/Bergstr.

TABLE OF CONTENTS

To

Margarita Hill Matlis

INTRODUCTION

The purpose of these notes is to present a structure theory for Artinian modules over a 1-dimensional, Noetherian, Cohen-Macaulay ring, and to show that many of the properties of such a ring may be derived from a knowledge of this theory. In fact, we shall be able to recapture many of the known results about 1-dimensional Cohen-Macaulay rings by these techniques, and find some new ones as well. Thus we have a new and unifying way of looking at these rings by examining a category of modules that previously has received scant attention.

My original aim was to study the category of Artinian divisible modules over a Noetherian domain of Krull dimension 1, and to see if there was a structure theory that would generalize the elementary theory of divisible modules over a Dedekind ring. I found that it was possible to apply the general theorems about integral domains developed in my paper, "Cotorsion Modules" [12] to yield striking results about the problems I was considering. After a time it became apparent that if divisible, torsion-free, torsion and cotorsion were all redefined in terms of the regular elements (that is, the elements that are not zero divisors) of an arbitrary commutative ring, then practically all of the tool theorems remained valid. Thereupon most of the striking results about divisible modules became true for an arbitrary 1-dimensional, Cohen-Macaulay ring (even though it had zero divisors). But perhaps the most surprising thing of all was the discovery that by using the structure theory of Artinian divisible modules it became possible to really understand and relate to each other the most important things about a 1-dimensional, local, Cohen-Macaulay ring; namely, the multiplicity, reduction number, latent multiplicities, latent residue degrees, and so forth. Theorems about Gorenstein rings, the existence of canonical ideals, analytic irreducibility, analytically unramified rings, and the analytic components are all seen to be theorems about the structure of Artinian divisible modules, and these theorems are all unified by this structure theory.

The structure of an Artinian module over a Dedekind ring is quite elementary. In the Dedekind case an Artinian module is a direct sum of a module of finite length and of a finite number of divisible modules of the form Z_{p_∞} for various primes p. Z_{p_∞} is a simple divisible module in the sense that it has no proper, nonzero divisible submodules. Since this decomposition of an Artinian module characterizes Dedekind rings, we can not hope to find anything as good for an arbitrary 1-dimensional Cohen-Macaulay ring.

The problem of finding the structure of an Artinian module is essentially a local problem, and for the sake of simplicity we shall assume for the rest of this introduction that R is a Noetherian, local, 1-dimensional Cohen-Macaulay ring. While we shall not be able to attain the precision that is possible for a Dedekind ring, we shall be able to present a satisfactory structure theory for Artinian R-modules that is far more complex than for Dedekind rings, but perhaps more interesting for that reason.

We shall show that an Artinian R-module is the sum of an R-module of finite length and of a divisible Artinian R-module. Every Artinian divisible R-module has a composition series of divisible submodules such that the factor modules are simple divisible R-modules in the sense described previously. We shall prove that there exists a Jordan-Hölder theorem for these composition series, but we shall have to replace the isomorphism of the factors of two composition series with the notion of equivalence.

Two modules are defined to be equivalent, if each is a homomorphic image of the other. Two Artinian, divisible R-modules A and B are equivalent if and only if $B \cong A/C$, where C is a finitely generated submodule of A. Thus, contrary to the usual practice of considering only finitely generated R-modules over a Noetherian ring, we actually go to the extreme of "throwing them away".

The Jordan-Hölder theorem for Artinian divisible R-modules gives us the concept of the divisible length of an Artinian R-module, a function that proves to be most useful. Since modulo its divisible submodule an Artinian R-module has finite length in the classical sense, we see that we have two numerical invariants to describe an Artinian R-module.

We prove that there are only a finite number of equivalence classes of simple divisible R-modules; and these are in one to one correspondence with the prime ideals of rank 0 of the completion of R, and hence with the analytic components of R. Each analytic component of R corresponds uniquely with one of the pseudo valuation rings containing R and contained in the full ring of quotients Q of R. (A ring is called a pseudo valuation ring if modulo its divisible submodule, it becomes an ordinary valuation ring.) If $V_1,\ldots,V_n$ are these pseudo valuation rings, then $Q/V_1,\ldots,Q/V_n$ are a complete set of representatives of the equivalence classes of simple divisible R-modules.

The set of isomorphism classes of R-modules equivalent to a given simple, divisible R-module is in one-to-one correspondence with the ideal class semi-group of the corresponding analytic component of R. And by analogy with Schur's Lemma we find that the endomorphism rings of these simple, divisible R-modules give us all of the integral extensions of this analytic component in its quotient field. Hence these endomorphism rings are complete, Noetherian, local domains of Krull dimension 1.

We define a semi-simple divisible R-module to be one that is equivalent to a finite direct sum of simple, divisible R-modules. We find that every Artinian divisible R-module is semi-simple if and only if R is analytically unramified. This result leads to a new proof of the known theorem that R is analytically unramified if and only if the integral closure of R is finitely generated.

While there is no semi-simple decomposition in general, there is, however, a primary decomposition for Artinian, divisible R-modules without any restrictions on R. We define an Artinian divisible R-module to be a P-primary divisible R-module if all of its composition factors correspond to the same fixed prime ideal P of rank 0 of the completion of R. We then see that every Artinian, divisible R-module is equivalent to a direct sum of uniquely determined P-primary divisible R-modules for the various primes P.

If Q is the full ring of quotients of R and K is the R-module Q/R, then K is an Artinian divisible R-module; and, in fact K contains the "genetic code" for all Artinian, divisible R-modules. A representative of each equivalence class of simple, divisible R-modules appears as a factor in a composition series for K, and the number of times a module equivalent to it appears as a factor in the composition series is equal to the latent multiplicity of the corresponding prime ideal of rank 0 of the completion of R.

A proper submodule A/R of K is a divisible R-module if and only if A is a strongly unramified ring extension of R. (If M is the maximal ideal of R, then a commutative ring $A \supset R$ is called strongly unramified over R if AM is the only regular maximal ideal of A and $A/(AM) \cong R/M$.) There exist only a finite number of maximal strongly unramified ring extensions of R in Q and these are called the closed components of R. We show that a subring of Q is a closed component of R if and only if it has an analytic component of R as its completion. Equivalently, A is a closed component of R if and only if A/R is the divisible submodule of V/R, where V is one of the pseudo valuation rings containing R and contained in Q. Modulo their divisible submodules the closed components of R are analytically irreducible Noetherian local domains of Krull dimension 1, and hence we have a way of passing to an analytically irreducible domain by using induction on the divisible length of K.

The point of view in writing these notes has been to make them as self contained as possible, and the reader is only assumed to possess an elementary knowledge of commutative ring theory and homological algebra. Most of the relevant definitions have been provided and proofs have been given of most of the general theorems needed as tools and preliminaries for this particular study. We have tried to save the reader the trouble of searching through the literature for the bits of machinery that have to be used.

The first two chapters of these notes present the necessary background information concerning h-divisible, cotorsion, and complete modules that we need in this investigation. This material is based on the paper "Cotorsion Modules" [12], but has been reworked so that it is applicable to arbitrary commutative rings. Of particular importance is the duality in Corollary 2.4 between torsion-free and complete modules on the one hand and torsion, h-divisible modules on the other. Much of the analysis is based on this result. We also develop here the properties of H, the completion of R in the R-topology. In the applications we make, the M-adic topology coincides with the R-topology and thus H is the ordinary M-adic completion of R. This coming together of the two topologies gives us the powerful additional tool of the duality between the Artinian R-modules and the Noetherian modules of the M-adic completion of R found in [9].

Chapter III deals with the theory of compatible ring extensions as enunciated in [16]. While elementary in nature, this forms the basis for the treatment of strongly unramified ring extensions later in the notes. Chapter IV shows how the problem of considering divisible modules over a 1-dimensional, Noetherian Cohen-Macaulay ring may be reduced to the local case. It also shows that for these rings a divisible module is h-divisible, and that if the ring has no nonzero nilpotent elements then the torsion submodule of a divisible module is a direct summand. This is a reworking of material in [10]. Theorems 4.5 and 4.6 collect together all of the necessary information concerning injective modules found in [9] and are presented without proofs.

Chapters V through XI present the theory of Artinian, divisible modules as outlined in this introduction, and most of it appears here in print for the first time. Chapters XII and XIII develop the theory of the multiplicity and reduction number of a 1-dimensional local Cohen-Macaulay ring and this is applied to the study of Gorenstein rings. A special role is played here by the concept of the first neighborhood ring, and some of Northcott's pioneering work is explained and carried further. Basically these two chapters are concerned with the study of Artinian modules that have finite length in the classical sense. The material here may be found in [15].

Chapter XIV combines the results of the preceding two chapters with those concerning Artinian divisible modules, and unites these two different streams of development.

Chapter XV considers the question of canonical ideals, and shows that the existence of such ideals is decided by the divisible structure of K.

An effort has been made throughout these notes to be scrupulous about providing attribution for those theorems not due originally to the author.

CHAPTER I

h-DIVISIBLE AND COTORSION MODULES

Throughout these notes R will be a commutative ring. An element of R that is not a zero divisor in R will be called a regular element of R. The set $\mathcal{S}$ of regular elements of R is multiplicatively closed. The ring $R_{\mathcal{S}}$ is called the full ring of quotients of R. We shall consistently use the notation $Q = R_{\mathcal{S}}$ and $K = Q/R$. We always assume that $Q \neq R$. If A is an R-module we shall denote the module $A_{\mathcal{S}}$ by the symbol A_Q.

In this chapter we shall review some elementary definitions and prove some basic theorems that we shall need as tools for what is to come later. Most of this material is a generalization of similar concepts for integral domains that may be found in [12].

Definitions: An R-module is said to be divisible if the multiplications on the module by the regular elements of R are all epimorphisms. Clearly a homomorphic image of a divisible R-module is divisible. On the other hand if A is a submodule of an R-module B, and if both A and B/A are divisible, then B is also divisible. The sum of two divisible submodules of a module is again divisible. Thus an R-module B has a unique largest divisible submodule d(B) that contains every divisible submodule of B. B is said to be reduced if it has no divisible submodules; that is, if $d(B) = 0$. Clearly, $B/d(B)$ is reduced.

The concept dual to divisible is that of torsion-free. An R-module is said to be torsion-free if the multiplications on the module by the regular elements of R are all monomorphisms. A submodule of a torsion-free R-module is torsion-free. If A is a submodule of an R-module B, and if both A and B/A are torsion-free, then B is also torsion-free. An R-module is said to be a torsion R-module if it has no nonzero torsion-free submodules. An R-module B has a unique largest torsion submodule t(B) that contains every torsion submodule of B, and B/t(B) is torsion-free. B is said to be torsion of bounded order if there is a regular element r of R such that rB = 0. It is easy to see that an R-module is both torsion-free and divisible if and only if it is a Q-module.

A more useful concept than divisible for the purposes of applying homological methods is that of h-divisible. An R-module is said to be h-divisible if it is a homomorphic image of a torsion-free and divisible R-module; that is, if it is an R-homomorphic image of a Q-module. If Q is a semi-simple ring (in particular, if R is an integral domain), then an R-module is h-divisible if and only if it is a homomorphic image of an injective R-module. It follows from the definition that an h-divisible R-module is divisible. We shall have something to say about the converse later.

A direct sum of h-divisible R-modules is again h-divisible. Consequently an R-module B contains a unique largest h-divisible submodule h(B) that contains every h-divisible submodule of B. B is said to be h-reduced, if it has no nonzero h-divisible submodules; that is, if h(B) = 0. Unfortunately, it is not true in general that B/h(B) is h-reduced. We shall examine presently some necessary and sufficient conditions for this desirable situation to occur.

Theorem 1.1. *If B is an R-module, then we have three fundamental exact sequences:*

(1) $0 \to B/t(B) \to Q \otimes_R B \to K \otimes_R B \to 0.$

(2) $0 \to \mathrm{Hom}_R(K,B) \to \mathrm{Hom}_R(Q,B) \to h(B) \to 0.$

(3) $0 \to B/h(B) \to \mathrm{Ext}^1_R(K,B) \to \mathrm{Ext}^1_R(Q,B) \to 0.$

Therefore, we have $t(B) \cong \mathrm{Tor}^R_1(K,B)$; *and B is a torsion R-module if and only if* $Q \otimes_R B = 0$. *Furthermore, B is h-reduced if and only if* $\mathrm{Hom}_R(Q,B) = 0$.

Proof. As our starting point we take the exact sequence:

(*) $$0 \to R \to Q \to K \to 0.$$

If we tensor this sequence with B and identify B with $R \otimes_R B$, we have an exact sequence:

$$0 \to \mathrm{Tor}_1^R(K,B) \to B \xrightarrow{\alpha} Q \otimes_R B \to K \otimes_R B \to 0$$

where $\alpha(x) = 1 \otimes x$ for $x \in B$. To establish (1) we have to show that $\ker \alpha = t(B)$. Because $Q \otimes_R B$ is a Q-module, it is torsion-free, and thus $t(B) \subset \ker \alpha$. Conversely, suppose that $x \in \ker \alpha$. Then the image of $Q \otimes_R Rx$ in $Q \otimes_R B$ is 0. But $Q \otimes_R \cdot$ is an exact functor and hence $Q \otimes_R Rx = 0$. Thus, there exists a regular element a of R such that $ax = 0$. Hence $x \in t(B)$ proving that $\ker \alpha = t(B)$, and that $\mathrm{Tor}_1^R(K,B) \cong t(B)$. We also see that $B = t(B)$ if and only if $Q \otimes_R B = 0$.

If we apply the functor $\mathrm{Hom}_R(\cdot,B)$ to exact sequence (*) and identify B with $\mathrm{Hom}_R(R,B)$, we obtain an exact sequence:

$$0 \to \mathrm{Hom}_R(K,B) \to \mathrm{Hom}_R(Q,B) \xrightarrow{\beta} B \to \mathrm{Ext}_R^1(K,B) \to \mathrm{Ext}_R^1(Q,B) \to 0,$$

where $\beta(f) = f(1)$ for $f \in \mathrm{Hom}_R(Q,B)$. Thus to establish (2) and (3) it is sufficient to prove that $\mathrm{Im}\,\beta = h(B)$.

Since $\mathrm{Hom}_R(Q,B)$ is a Q-module, it is torsion-free and divisible and thus $\mathrm{Im}\,\beta \subset h(B)$. On the other hand, there is a Q-module V mapping onto h(B). Let $x \in h(B)$ and choose an element $y \in V$ that maps onto x. Since $Qy \subset V$, there is an R-homomorphism $f: Q \to h(B)$ such that $f(1) = x$. Therefore, $x \in \mathrm{Im}\,\beta$ and hence $\mathrm{Im}\,\beta = h(B)$. It is now clear that B is h-reduced if and only if $\mathrm{Hom}_R(Q,B) = 0$.

Corollary 1.2. *Let B be a torsion R-module. Then the natural map:*

$$\varphi : K \otimes_R \mathrm{Hom}_R(K,B) \to h(B)$$

defined by $\varphi(x \otimes f) = f(x)$ *for* $x \in K$ *and* $f \in \mathrm{Hom}_R(K,B)$ *is an iso-*

morphism.

Proof. By Theorem 1.1, $K \otimes_R \mathrm{Hom}_R(K,B)$ is an R-homomorphic image of the Q-module $Q \otimes_R \mathrm{Hom}_R(K,B)$ and hence $K \otimes_R \mathrm{Hom}_R(K,B)$ is h-divisible. Thus Im φ is h-divisible, and hence Im $\varphi \subset h(B)$. On the other hand, let $y \in h(B)$. Then by Theorem 1.1 there is an R-homomorphism $\eta : Q \rightarrow h(B)$ such that $\eta(1) = y$. If $S = \ker \eta$, then Q/S is a torsion R-module, and hence there is a regular element a in $S \cap R$. Define $\nu : Q \rightarrow h(B)$ by $\nu(z) = \eta(az)$ for all $z \in Q$. Then $\nu(a^{-1}) = y$ and $\nu(1) = 0$. Therefore, $R \subset \ker \nu$ and we have an induced R-homomorphism $f : K \rightarrow h(B)$ such that $y \in f(K)$. Hence we see that Im $\varphi = h(B)$.

Every element of $K \otimes_R \mathrm{Hom}_R(K,B)$ can be written in the form $x \otimes f$, where $f \in \mathrm{Hom}_R(K,B)$ and $x = a^{-1} + R$ for some regular element $a \in R$. Suppose that $x \otimes f \in \ker \varphi$; that is $f(x) = 0$. Define $g \in \mathrm{Hom}_R(K,B)$ as follows: if $y \in K$, then since K is divisible there is an element $z \in K$ such that $y = az$, and we let $g(y) = f(z)$. If w is any element of K such that $aw = 0$, then $w \in Rx$ and hence $f(w) = 0$. This shows that g is a well-defined R-homomorphism. Clearly $ag = f$, and hence $f \otimes x = ag \otimes x = g \otimes ax = g \otimes 0 = 0$. Thus φ is an isomorphism.

Corollary 1.3. Let B be a torsion R-module. Then we have the following:

(1) If $x \in B$, then $x \in h(B)$ if and only if there exists $f \in \mathrm{Hom}_R(K,B)$ such that $x \in f(K)$.

(2) B is h-reduced if and only if $\mathrm{Hom}_R(K,B) = 0$.

(3) If A and C are h-divisible submodules of B, then A = C if and only if $\mathrm{Hom}_R(K,A) = \mathrm{Hom}_R(K,C)$.

Proof. This corollary follows immediately from Corollary 1.2.

Theorem 1.4. Assume that Q is a semi-simple ring. If B is an h-divisible R-module, then t(B) is a direct summand of B.

Proof. By Zorn's Lemma there exists a maximal torsion-free and divisible submodule C of B. Then B/C has no nonzero torsion-free and divisible submodules, and the torsion submodule of B/C is $(t(B) + C)/C$. Since $t(B) \cap C = 0$, we may assume without loss of generality that B has no nonzero torsion-free and divisible submodules, and prove that B is a torsion R-module.

Let x be a nonzero element of B. Then by Theorem 1.1, there is an R-homomorphism $f : Q \to B$ such that $f(1) = x$. If $S = \operatorname{Ker} f$, then Q/S is isomorphic to a submodule of B. Now the torsion submodule of Q/S is S_Q/S, and S_Q is an ideal of Q. Since Q is a semisimple ring, S_Q is a direct summand of Q, and thus S_Q/S is a direct summand of Q/S. Its complementary summand in Q/S is a torsion-free and divisible submodule of B, and hence is equal to 0. Therefore Q/S is a torsion R-module, and hence x is a torsion element of B. We thus have shown that B is a torsion R-module, and this completes the proof of the theorem.

Definition. An R-module C is said to be a cotorsion R-module if $\operatorname{Hom}_R(Q,C) = 0$ and $\operatorname{Ext}^1_R(Q,C) = 0$. Thus, in particular, a cotorsion module is h-reduced. If B is a torsion R-module of bounded order, then $\operatorname{Ext}^n_R(Q,B)$ is both torsion and torsion-free, and hence $\operatorname{Ext}^n_R(Q,B) = 0$ for all $n \geq 0$. Thus a torsion R-module of bounded order is a cotorsion R-module.

If C is a cotorsion R-module, and A is any Q-module, then $\operatorname{Hom}_R(A,C) = 0 = \operatorname{Ext}^1_R(A,C)$. For we have $\operatorname{Hom}_R(A,C) = 0$ because C is h-reduced. Now we can write $A = F/P$, where F is a free Q-module and P is a Q-submodule of F. We then have $\operatorname{Ext}^1_R(A,C) \cong \operatorname{Hom}_R(P,C) = 0$, establishing our assertion.

Theorem 1.5. *Let $0 \to A \to B \to C \to 0$ be an exact sequence of R-modules. Then we have the following:*

(1) *If A and C are cotorsion, then B is also cotorsion.*

(2) If B is cotorsion, then A is cotorsion if and only if C is h-reduced.

Proof. This theorem is an immediate consequence of the long exact sequence obtained by applying the functor $\mathrm{Ext}_R(Q,\cdot)$ to the given exact sequence.

Theorem 1.6. Let B and C be R-modules. If B is torsion, or if C is cotorsion, then $\mathrm{Hom}_R(B,C)$ is cotorsion.

Proof. We have the canonical duality isomorphism $\mathrm{Hom}_R(Q,\mathrm{Hom}_R(B,C)) \cong \mathrm{Hom}_R(Q \otimes_R B,C)$ by [3, Ch. II, Prop. 5.2]. Thus if B is torsion, or C is cotorsion, then $\mathrm{Hom}_R(B,C)$ is h-reduced. Suppose that B is a torsion R-module, and that E is an injective R-module containing C. Then we have an exact sequence of h-reduced R-modules:

$$0 \to \mathrm{Hom}_R(B,C) \to \mathrm{Hom}_R(B,E) \to \mathrm{Hom}_R(B,E/C).$$

We have $\mathrm{Ext}_R^n(Q,\mathrm{Hom}_R(B,E)) \cong \mathrm{Hom}_R(\mathrm{Tor}_n^R(Q,B),E) = 0$ for all $n \geq 0$ by [3, Ch. VI, Prop. 5.1]. Hence by Theorem 1.5, $\mathrm{Hom}_R(B,C)$ is a cotorsion R-module. Suppose that C is a cotorsion R-module, and let F be a free R-module mapping onto B with kernel A. Then we have an exact sequence of h-reduced R-modules.

$$0 \to \mathrm{Hom}_R(B,C) \to \mathrm{Hom}_R(F,C) \to \mathrm{Hom}_R(A,C).$$

$\mathrm{Hom}_R(F,C)$ is isomorphic to a direct product of copies of C and thus is a cotorsion R-module. Hence $\mathrm{Hom}_R(B,C)$ is cotorsion by Theorem 1.5.

Definition. The homological dimension of an R-module A (abbreviated $\mathrm{hd}_R A$) is defined to be the smallest integer n such that $\mathrm{Ext}_R^m(A,B) = 0$ for all $m \geq n+1$ and all R-modules B. (If such an integer n does not exist, then $\mathrm{hd}_R A = \infty$.) It is easy to see that $\mathrm{hd}_R A \leq 1$ if and only if $\mathrm{Ext}_R^1(A,B) = 0$ for every R-module B that is a homomorphic image of an injective R-module.

Theorem 1.7. The following statements are equivalent:

(1) $hd_R Q = 1$

(2) If $0 \to A \to B \to C \to 0$ is an exact sequence of R-modules and A and B are cotorsion, then C is cotorsion.

(3) $Ext_R^1(K,B)$ is a cotorsion R-module for all R-modules B.

(4) $Ext_R^1(K,B) = 0$ for all h-divisible R-modules B.

(5) $Ext_R^1(K,B) = 0$ for all torsion, h-divisible R-modules B

(6) $hd_R K = 1$.

Proof. (1) ⟹ (2). We see from Theorem 1.5 that C is an h-reduced R-module. Now we have an exact sequence:

$$Ext_R^1(Q,B) \to Ext_R^1(Q,C) \to Ext_R^2(Q,A).$$

Since B is a cotorsion R-module and $hd_R Q = 1$, the end terms of this sequence are zero. Thus $Ext_R^1(Q,C) = 0$, and hence C is a cotorsion R-module.

(2) ⟹ (3). Let B be an R-module and E an injective R-module containing B. Then we have an exact sequence:

$$0 \to Hom_R(K,B) \to Hom_R(K,E) \xrightarrow{\alpha} Hom_R(K,E/B) \to Ext_R^1(K,B) \to 0$$

By Theorem 1.6 every term in this sequence, except possibly $Ext_R^1(K,B)$ is a cotorsion R-module. Hence by (2), Im α is a cotorsion R-module. But then by (2) again, $Ext_R^1(K,B)$ is also a cotorsion R-module.

(3) ⟹ (4). If B is an h-divisible R-module, then by Theorem 1.1 (3) we have $Ext_R^1(K,B) \cong Ext_R^1(Q,B)$. Thus $Ext_R^1(K,B)$ is torsion-free and divisible. Since it is cotorsion by (3), we see that $Ext_R^1(K,B) = 0$.

(4) ⟹ (5). This is a trivial assertion.

(5) ⟹ (6). Let B be a homomorphic image of an injective R-

module. It is sufficient to prove that $\operatorname{Ext}_R^1(K,B) = 0$. It is easy to see that an injective R-module is h-divisible, and hence every homomorphic image of an injective R-module is h-divisible. We have an exact sequence:

$$\operatorname{Hom}_R(K,B/t(B)) \to \operatorname{Ext}_R^1(K,t(B)) \to \operatorname{Ext}_R^1(K,B) \to \operatorname{Ext}_R^1(K,B/t(B)).$$

Clearly $\operatorname{Hom}_R(K,B/t(B)) = 0$. On the other hand, since $B/t(B)$ is a Q-module, we have by [3, Ch. VI, Prop. 4.1.3] that $\operatorname{Ext}_R^1(K,B/t(B)) \cong \operatorname{Ext}_Q^1(Q \otimes_R K,B/t(B))$. But $Q \otimes_R K = 0$, and thus $\operatorname{Ext}_R^1(K,B/t(B)) = 0$. Therefore, from the previous exact sequence we see that $\operatorname{Ext}_R^1(K,t(B)) \cong \operatorname{Ext}_R^1(K,B)$. Thus by (5) it is sufficient to prove that $t(B)$ is h-divisible.

Let $x \in t(B)$; then by Theorem 1.1 there is an R-homomorphism $f : Q \to B$ such that $f(1) = x$. If $S = \operatorname{Ker} f$, then S contains a regular element of R and hence Q/S is a torsion R-module. Therefore $\operatorname{Im} f \subset t(B)$, and from this it follows that $t(B)$ is h-divisible.

(6) $\Longrightarrow$ (1). This assertion is obvious.

<u>Lemma 1.8. The following two statements are equivalent:</u>

(1) <u>$B/h(B)$ is h-reduced for every R-module B.</u>

(2) <u>If $0 \to A \to B \to C \to 0$ is an exact sequence of R-modules such that A and C are h-divisible, then B is also h-divisible.</u>

<u>Proof</u>. (1) $\Longrightarrow$ (2). Let A be a submodule of an R-module B, and suppose that A and B/A are h-divisible. Then $A \subset h(B)$, and hence $B/h(B)$ is a homomorphic image of B/A. Thus $B/h(B)$ is h-divisible. Since $B/h(B)$ is h-reduced by (1), we see that $B/h(B) = 0$. Therefore B is h-divisible.

(2) $\Longrightarrow$ (1). Let B be an R-module and A a submodule of B containing $h(B)$. If $A/h(B)$ is h-divisible, then A is h-divisible by (2). But then $A = h(B)$, showing that $B/h(B)$ is h-reduced.

<u>Theorem 1.9. If $\operatorname{hd}_R Q = 1$, then $B/h(B)$ is h-reduced for every</u>

R-module B. On the other hand, if Q is a semi-simple ring, then the two statements are equivalent.

Proof. If $hd_R Q = 1$, then by Theorem 1.7, $Ext_R^1(K,B)$ is a cotorsion R-module. Since $B/h(B)$ is isomorphic to a submodule of $Ext_R^1(K,B)$ by Theorem 1.1 (3), we see that $B/h(B)$ is h-reduced.

On the other hand let us assume that Q is a semi-simple ring and that $B/h(B)$ is h-reduced for every R-module B. Let A be a torsion h-divisible R-module. To prove that $hd_R Q = 1$ it is sufficient by Theorem 1.7 to prove that $Ext_R^1(K,A) = 0$. By Theorem 1.1 (3) we have $Ext_R^1(K,A) \cong Ext_R^1(Q,A)$. Hence it is sufficient to prove that $Ext_R^1(Q,A) = 0$.

Let us consider an exact sequence of the form:

$$0 \to A \to C \to Q \to 0.$$

Then A is the torsion submodule of C. By Lemma 1.8, C is an h-divisible R-module; and, since Q is a semi-simple ring, we see from Theorem 1.4 that A is a direct summand of C. This proves that $Ext_R^1(Q,A) = 0$, and hence $hd_R Q = 1$.

Remarks. Hamsher [4] has proved for an integral domain R that $hd_R Q = 1$ if and only if every divisible R-module is h-divisible. Therefore, because of Theorem 1.4, both of these conditions are equivalent to the condition that the torsion submodule of a divisible module is a direct summand. I suspect that these results hold true even if we only assume that Q is a semi-simple ring instead of a field. However, we shall not need these facts, and thus we shall not reproduce here the proof of Hamsher's result.

Theorem 1.10. Let A be an R-module and C a cotorsion R-module containing A such that C/A is torsion-free and divisible. Then the following statements are true:

(1) If D is a cotorsion R-module, then every R-homomorphism from A into D can be lifted uniquely to an R-homomorphism of C into D.

(2) *If* D *is a cotorsion* R-*module containing* A, *and* D/A *is torsion-free and divisible, then* D *is isomorphic to* C.

Proof. We have an exact sequence:

$$\mathrm{Hom}_R(C/A,D) \to \mathrm{Hom}_R(C,D) \to \mathrm{Hom}_R(A,D) \to \mathrm{Ext}^1_R(C/A,D).$$

The end terms of this sequence are 0, since D is a cotorsion R-module. Thus $\mathrm{Hom}_R(C,D) \cong \mathrm{Hom}_R(A,D)$. The first statement of the theorem follows easily from this, and the second statement follows easily from the first.

CHAPTER II

COMPLETIONS

Definitions: We shall say that an ideal of R is a regular ideal if it contains a regular element of R. If A is an R-module, we define a topology on A, called the R-topology of A, by taking the submodules of the form IA, where I is a regular ideal of R, as a system of neighborhoods of 0 in A. The submodules of the form rA, where r is a regular element of R, give the same topology. This is a uniform topology, and hence every R-module has a unique completion denoted by $\tilde{A}$. It is standard point set topology that $\tilde{A}$ is equal to the inverse limit of the R-modules A/rA:

$$\tilde{A} = \varprojlim A/rA.$$

Let C be the direct product of the R-modules A/rA, where r ranges over the regular elements of R. Then $\tilde{A}$ is a submodule of C. If we represent an element $\tilde{x}$ of C in the form $\langle x_r + rA \rangle$, where $x_r \in A$, then $\tilde{x} \in \tilde{A}$ if and only if $x_r - x_{rs} \in rA$ for every regular r and s in R. We have a natural R-homomorphism $\eta : A \to \tilde{A}$ given by $\eta(x) = \langle x + rA \rangle$ for every $x \in A$. The kernel of η is equal to $\bigcap_r rA$, where r ranges over the regular elements of R.

Theorem 2.1. Let A be an R-module whose torsion submodule has bounded order. Then

(1) Ker η is the divisible submodule of A and is torsion-free.

(2) $\tilde{A}$ is a cotorsion R-module.

(3) $t(\tilde{A}) \cong t(A)$, and thus if A is torsion-free, $\tilde{A}$ is also torsion-free.

(4) $\tilde{A}/\eta(A)$ is torsion-free and divisible and $\tilde{A}/\eta(A) \cong \mathrm{Ext}_R^1(Q,A)$.

Proof. (1) It is clear that ker $\eta = \bigcap rA$, where r ranges over all regular elements of R. Since t(A) has bounded order, $\bigcap rA$ is the divisible submodule of A and is torsion-free.

(2) For each regular element $s \in R$ define $\varphi_s : C \to C$ by $\varphi_s(\langle z_r + rA\rangle) = \langle z_r - z_{rs} + rA\rangle$ (where $z_r \in A$). Then $\widetilde{A} = \bigcap_s \ker \varphi_s$, where s ranges over all regular elements of R. Thus we have a monomorphism of $C/\widetilde{A}$ into a direct product of copies of C. Clearly any direct product of copies of C is a cotorsion R-module. Thus by Theorem 1.5, $\widetilde{A}$ is a cotorsion R-module.

(3) Let $\widetilde{x} = \langle x_r + rA\rangle$ be an element of $t(\widetilde{A})$. Then there is a fixed regular element $s \in R$ such that $sx_r \in rA$ for every regular element $r \in R$. Hence $x_{rs} \in rA + t(A)$; and since $x_r - x_{rs} \in rA$, we can assume without loss of generality that $x_r \in t(A)$ for every r. Let v be a regular element of R such that $vt(A) = 0$. Then $x_{rv} - x_v \in vA \cap t(A) = vt(A) = 0$. Thus $x_r - x_v = (x_r - x_{rv}) + (x_{rv} - x_v) = x_r - x_{rv} \in rA$. Therefore, $\widetilde{x} = \eta(x_v)$ and hence $t(\widetilde{A}) = t(\eta(A))$. Because Ker η is torsion-free and divisible, $t(\eta(A)) \cong t(A)$.

(4) Let $\widetilde{x} = \langle x_r + rA\rangle \in \widetilde{A}$ and suppose that $s\widetilde{x} = \eta(x)$ for some regular element $s \in R$ and $x \in A$. Then $sx_r - x \in rA$ for every regular element $r \in R$. Hence $x = sy$ for some $y \in A$, and $\widetilde{x} - \eta(y) \in t(\widetilde{A}) \subset \eta(A)$. Therefore, $\widetilde{x} \in \eta(A)$; and we have proved that $\widetilde{A}/\eta(A)$ is torsion-free.

We shall next prove that $\widetilde{A}/\eta(A)$ is divisible. Let s be a regular element of R and $\widetilde{x} = \langle x_r + rA\rangle$ an element of $\widetilde{A}$. Let v be a regular element of R such that $vt(A) = 0$, and let $u = vs$. For each regular element $r \in R$ there is an element $y_r \in A$ such that $x_{ru} - x_u = uy_r$. For any regular element $q \in R$, $uy_r - uy_{rq} \in urA$. Hence $y_r - y_{rq} \in rA + t(A)$, and thus $vy_r - vy_{rq} \in rA$. Let $\widetilde{y} = \langle vy_r + rA\rangle$; then $\widetilde{y} \in \widetilde{A}$ and $\widetilde{x} - s\widetilde{y} = \langle x_{ru} - uy_r + rA\rangle = \langle x_u + rA\rangle = \eta(x_u) \in \eta(A)$. Therefore $\widetilde{A}/\eta(A)$ is divisible.

We have an exact sequence:

$$\mathrm{Hom}_R(Q,\widetilde{A}) \to \mathrm{Hom}_R(Q,\widetilde{A}/\eta(A)) \to \mathrm{Ext}^1_R(Q,\eta(A)) \to \mathrm{Ext}^1_R(Q,\widetilde{A}).$$

The end terms of this sequence are zero because $\widetilde{A}$ is cotorsion. Hence $\mathrm{Hom}_R(Q,\widetilde{A}/\eta(A)) \cong \mathrm{Ext}^1_R(Q,\eta(A))$. But since $\widetilde{A}/\eta(A)$ is torsion-free and divisible, we have $\mathrm{Hom}_R(Q,\tilde{A}/\eta(A)) \cong \tilde{A}/\eta(A)$. Thus $\tilde{A}/\eta(A)$ $\cong \mathrm{Ext}_R^1(Q,\eta(A))$. Since Ker η is a Q-module, we have $\mathrm{Ext}_R^n(Q,\mathrm{Ker}\ \eta)$ $\cong \mathrm{Ext}_Q^n(Q,\mathrm{Ker}\ \eta) = 0$ for all $n > 0$ by [3, Ch. VI, Prop. 4.1.3]. It follows from this that $\mathrm{Ext}_R^1(Q,\eta(A)) \cong \mathrm{Ext}_R^1(Q,A)$.

Definition. Let A be an R-module, and $a \in A$. We define $a^* : K \to K \otimes_R A$ by $a^*(x) = x \otimes a$ for every $x \in K$. We can then define an R-homomorphism $\lambda : A \to \mathrm{Hom}_R(K,K \otimes_R A)$ by $\lambda(a) = a^*$.

Theorem 2.2. *Let A be a torsion-free R-module. Then the kernel of $\lambda : A \to \mathrm{Hom}_R(K,K \otimes_R A)$ is the divisible submodule of A; and there is an isomorphism $\tilde{\lambda}$ of $\tilde{A}$ onto $\mathrm{Hom}_R(K,K \otimes_R A)$ such that $\lambda = \tilde{\lambda}\eta$.*

Proof. Every element of $K \otimes_R A$ can be written in the form $x \otimes a$, where $a \in A$ and $x = r^{-1} + R$ for some regular element $r \in R$. If $x \otimes a$ is of this form, then $x \otimes a = 0$ if and only if $a \in rA$. For if $x \otimes a = 0$, then in $Q \otimes_R A$ we have $r^{-1} \otimes a = 1 \otimes b = r^{-1} \otimes rb$ for some $b \in A$, and hence $1 \otimes (a-rb) = 0$. Therefore, $a - rb = 0$ by Theorem 1.1. It is now clear that Ker $\lambda = \bigcap rA$, where r ranges over the regular elements of R, and hence Ker λ is the divisible submodule of A.

Let $\tilde{a} = \langle a_r + rA \rangle$ be an element of $\tilde{A}$. We define an R-homomorphism $\tilde{a}^* : K \to K \otimes_R A$ by $\tilde{a}^*(x) = x \otimes a_r$ if $x \in K$ and $rx = 0$ for some regular element r of R. It is not hard to verify that $\tilde{a}^*$ is a well-defined R-homomorphism. We can then define an R-homomorphism $\tilde{\lambda} : \tilde{A} \to \mathrm{Hom}_R(K,K \otimes_R A)$ by $\tilde{\lambda}(\tilde{a}) = \tilde{a}^*$. It is clear that $\lambda = \tilde{\lambda}\eta$.

Suppose that $\tilde{a} = \langle a_r + rA \rangle$ is an element of Ker $\tilde{\lambda}$. Then by the first paragraph the proof, $a_r \in rA$ for every regular $r \in R$, and hence $\tilde{a} = 0$. Thus $\tilde{\lambda}$ is a monomorphism.

Let $f \in \mathrm{Hom}_R(K,K \otimes_R A)$; then if r is a regular element of R, there is an element $a_r \in A$ such that $f(r^{-1}+R) = (r^{-1}+R) \otimes a_r$. For $f(r^{-1}+R) = (c^{-1}+R) \otimes b$ where $b \in A$ and c is a regular element of R. Then $0 = (c^{-1}+R) \otimes rb$, and hence by the first paragraph of the proof there is an element $a_r \in A$ such that $rb = ca_r$. Thus $(r^{-1}+R) \otimes a_r = (c^{-1}r^{-1}+R) \otimes ca_r = (c^{-1}r^{-1}+R) \otimes rb = (c^{-1}+R) \otimes b = f(r^{-1}+R)$. Now for any regular element $s \in R$ we have: $(r^{-1}+R) \otimes (a_r - a_{rs}) = f(r^{-1}+R) - sf(s^{-1}r^{-1}+R) = 0$. Hence $a_r - a_{rs} \in rA$, and thus $\tilde{a} = \langle a_r + rA \rangle$ is an element of $\tilde{A}$. Clearly $\tilde{\lambda}(\tilde{a}) = f$, and we have shown that $\tilde{\lambda}$ is an isomorphism.

Corollary 2.3. *Let A be a torsion-free R-module. Then the following statements are equivalent:*

(1) *A is a cotorsion R-module.*

(2) *A is complete in the R-topology.*

(3) $A \cong \mathrm{Hom}_R(K, K \otimes_R A)$.

(4) $A \cong \mathrm{Hom}_R(K,B)$ *for some R-module B.*

Proof. (1) ==> (2). By Theorem 2.1, $\tilde{A}$ is cotorsion, and $\tilde{A}/A$ is torsion-free and divisible. But $\tilde{A}/A$ is a reduced R-module by Theorem 1.5. Hence $A = \tilde{A}$, and thus A is complete in the R-topology.

(2) ==> (3). This is an immediate consequence of Theorem 2.2.

(3) ==> (4). This is a trivial assertion.

(4) ==> (1). This follows directly from Theorem 1.6.

Corollary 2.4. (Duality). *The functor* $\mathrm{Hom}_R(K,\cdot)$ *is a natural equivalence from the category* $\mathcal{A}$ *of torsion-free, cotorsion R-modules onto the category* $\mathcal{B}$ *of torsion h-divisible modules; and its inverse and adjoint is given by* $K \otimes_R \cdot$.

If A is in $\mathcal{A}$ *and B is in* $\mathcal{B}$, *then*

(1) $\mathrm{Hom}_R(K \otimes_R A, B) \cong \mathrm{Hom}_R(A, \mathrm{Hom}_R(K,B))$

(2) $\mathrm{Hom}_R(B, K \otimes_R A) \cong \mathrm{Hom}_R(\mathrm{Hom}_R(K,B), A)$.

Proof. The first statement of the theorem is a consequence of Corollary 1.2 and Corollary 2.3. The first isomorphism is the canonical duality isomorphism [3, Ch. II, Prop. 5.2]. To prove the second isomorphism, let $C = \mathrm{Hom}_R(K,B)$. Then $B \cong K \otimes_R C$ and

$A \cong \mathrm{Hom}_R(K, K \otimes_R A)$. Thus the second isomorphism reduces to the first.

Definitions. We shall let H denote the completion of R in the R-topology. If we identify $K \otimes_R R$ with K, then by Theorem 2.2, H is naturally isomorphic to $\mathrm{Hom}_R(K,K)$, and the mapping λ of R into H of Theorem 2.2 sends an element of R into multiplication by that element on K.

Corollary 2.5.

(1) H is a commutative ring, and is a torsion-free R-module.

(2) We have an exact sequence

$$0 \to d(R) \to R \xrightarrow{\lambda} H \to \mathrm{Ext}^1_R(Q,R) \to 0$$

where $d(R)$ is the divisible submodule of R and λ is a ring homomorphism.

(3) $\mathrm{Tor}^R_1(H,B) = 0$ for every torsion R-module B and $\mathrm{Ext}^1_R(H,C) = 0$ for every cotorsion R-module C.

Proof. Let f be an element of H, let r be a regular element of R and let $y = r^{-1} + R$ in K.

Since $rf(y) = 0$, it follows that $f(y) \in Ry$, and hence $f(y) = sy$ for some $s \in R$ (depending on y). If g is another element of H, then $g(y) = vy$ for some $v \in R$. Thus $g(f(y)) = g(sy) = svy = f(vy) = f(g(y))$. Hence $gf = fg$, showing that H is a commutative ring.

Since $H = \mathrm{Hom}_R(K,K)$, H is a torsion-free R-module.

(2) The exact sequence is a direct consequence of Theorems 2.1 and 2.2.

(3) This is a fairly easy consequence of applying the functors $\cdot \otimes_R B$ and $\mathrm{Hom}_R(\cdot, C)$ to the exact sequence of (2).

Remarks. As we shall see in Theorem 2.7, if I is a regular ideal of R, then $H/HI \cong H \otimes_R R/I \cong R/I$. It follows that the completion of H in the R-topology is equal to the completion of R in the R-topology; that is, H is complete in the R-topology.

Corollary 2.6. *Assume that* R *is a reduced* R-*module. Then:*

(1) *If* Q *is a semi-simple ring, then* H *is a faithfully flat* R-*module.*

(2) *If* H *is a faithfully flat* R-*module and* A *is any* R-*module, then* $H \otimes_R A$ *is a finitely generated projective* H-*module if and only if* A *is a finitely generated projective* R-*module.*

(3) *If* R *is a Noetherian local ring and* $hd_R Q = 1$, *then* H *is a faithfully flat* R-*module and* $H \otimes_R A$ *is the completion of* A *for every finitely generated* R-*module* A.

Proof. (1) Since Q is a semi-simple ring, the Q-module $\mathrm{Ext}^1_R(Q,R)$ is a projective Q-module. Because Q is flat over R, it follows that $\mathrm{Ext}^1_R(Q,R)$ is a flat R-module. Hence exact sequence (2) of Corollary 2.5 shows that H is a faithfully flat R-module (see [2]).

(2) This is a property of faithfully flat ring extensions of R (see [2]).

(3) If A is a finitely generated R-module, we have an exact sequence: $F_1 \to F_0 \to A \to 0$, where F_1 and F_0 are finitely generated free R-modules. Hence we have an exact sequence:

$$H \otimes_R F_1 \xrightarrow{\alpha} H \otimes_R F_0 \xrightarrow{\beta} H \otimes_R A \to 0,$$

and $H \otimes_R F_1$ and $H \otimes_R F_0$ are finitely generated free H-modules. Therefore Im α is an h-reduced R-module, and hence Ker α is a cotorsion R-module by Theorem 1.5. But since $hd_R Q = 1$, Im α is a cotorsion R-module by Theorem 1.7. Because Im α = Ker β, it follows from another application of Theorem 1.7 that $H \otimes_R A$ is a cotorsion R-module.

We have an exact sequence:

$$0 \to \mathrm{Tor}_1^R(H,A) \to \mathrm{Tor}_1^R(H/R,A) \xrightarrow{\delta} A \to H \otimes_R A \to (H/R) \otimes_R A \to 0.$$

Because A is finitely generated, it follows from the Nakayama Lemma that A is a reduced R-module. However $\mathrm{Tor}_1^R(H/R,A)$ is divisible because H/R is a Q-module. Therefore $\delta = 0$ and we have an isomorphism $\mathrm{Tor}_1^R(H,A) \cong \mathrm{Tor}_1^R(H/R,A)$, and an imbedding $A \subset H \otimes_R A$ such that $(H \otimes_R A)/A$ is torsion-free and divisible. If $\tilde{A}$ is the completion of A, then $\tilde{A}$ is a cotorsion R-module and $\tilde{A}/A$ is torsion-free and divisible by Theorem 2.1. Hence by Theorem 1.10, $\tilde{A}$ is isomorphic to $H \otimes_R A$.

If I is an ideal of R, then we have an exact sequence:

$$0 \to \mathrm{Tor}_1^R(H,R/I) \to H \otimes_R I \to H \to H \otimes_R R/I \to 0.$$

But we have shown that $\mathrm{Tor}_1^R(H,R/I) \cong \mathrm{Tor}_1^R(H/R,R/I)$ is torsion-free and divisible and $H \otimes_R I$ is cotorsion. Therefore, $\mathrm{Tor}_1^R(H,R/I) = 0 = \mathrm{Tor}_1^R(H/R,R/I)$. It follows that H and H/R are flat R-modules. Thus, as is well known (see [2]), H is a faithfully flat R-module.

Theorem 2.7. (1) *If T is a torsion R-module, then $T \cong H \otimes_R T$. Thus T has the structure of an H-module, and this structure is unique. Every R-submodule of T is also an H-submodule of T. If A is any H-module, then $\mathrm{Hom}_R(T,A) = \mathrm{Hom}_H(T,A)$.*

(2) *If C is a cotorsion R-module, then $C \cong \mathrm{Hom}_R(H,C)$. Thus C has the structure of an H-module, and this structure is unique. If A is any H-module, then $\mathrm{Hom}_R(A,C) = \mathrm{Hom}_H(A,C)$.*

Proof. (1) Because $d(R) \otimes_R T = 0 = \mathrm{Ext}_R^1(Q,R) \otimes_R T$, we see that if we tensor the exact sequence of Corollary 2.5 with T, we obtain an isomorphism: $\theta : T \to H \otimes_R T$, where θ is defined by $\theta(x) = 1 \otimes x$ for all $x \in T$. Thus T has an H-module structure extending that of R defined by $fx = \theta^{-1}(f \otimes x)$ for all $f \in H$ and $x \in T$.

Suppose that there is another H-module structure on T extending that of R, and let us denote it by $f \circ x$. There is a regular element

$r \in R$ (depending on x) such that $rx = 0$. Because $H/\lambda(R)$ is divisible, there are elements $g \in H$ and $s \in R$ such that $f = rg + \lambda(s)$. Therefore, $fx = sx = f \circ x$; and hence the two H-structure are identical. The fact that $fx = sx$ also shows that every R-submodule of T is also an H-submodule of T.

If φ is an R-homomorphism of T into an H-module A, then $\varphi(fx) = \varphi(sx) = s\varphi(x) = f\varphi(x) - g\varphi(rx) = f\varphi(x)$. Thus we see that φ is an H-homomorphism, and hence $\mathrm{Hom}_R(T,A) = \mathrm{Hom}_H(T,A)$.

(2) We have an exact sequence:

$$\mathrm{Hom}_R(H/\lambda(R),C) \to \mathrm{Hom}_R(H,C) \xrightarrow{\nu} C \to \mathrm{Ext}^1_R(H/\lambda(R),C)$$

where ν is defined by $\nu(\varphi) = \varphi(1)$ for all $\varphi \in \mathrm{Hom}_R(H,C)$. Since C is a cotorsion R-module, and $H/\lambda(R)$ is torsion-free and divisible by Corollary 2.5, the end modules of this sequence are zero, and thus ν is an isomorphism. Therefore, C has the structure of an H-module defined by $fx = \varphi(f)$, where $f \in H$, $x \in C$ and φ is an element of $\mathrm{Hom}_R(H,C)$ such that $\varphi(1) = x$.

Suppose that C has another H-module structure and let us denote it by $f \bullet x$. For a fixed $x \in C$, define $\eta : H \to C$ by $\eta(f) = f \bullet x$. Then $\nu(\eta) = \eta(1) = x = \varphi(1) = \nu(\varphi)$, and since ν is an isomorphism, we have $\eta = \varphi$. Therefore $f \bullet x = fx$, and we see that the H-structure on C is unique.

Let A be any H-module; then since $C \cong \mathrm{Hom}_R(H,C)$ we can utilize the canonical duality isomorphism of [3, Ch. II, Prop. 5.2] and obtain:

$$\mathrm{Hom}_H(A,C) \cong \mathrm{Hom}_H(A,\mathrm{Hom}_R(H,C)) \cong \mathrm{Hom}_R(H \otimes_H A,C) \cong \mathrm{Hom}_R(A,C).$$

<u>Definitions:</u> A commutative ring is called a <u>quasi-local ring</u> if it has only one maximal ideal. A Noetherian, quasi-local ring is called a <u>local ring</u>, and a Noetherian ring with only a finite number of maximal ideals is called a <u>semi-local ring</u>.

Theorem 2.8. *There is a one-to-one correspondence between the set of regular ideals* I *of* R *and the set of regular ideals* J *of* H *such that* $\lambda^{-1}(J)$ *is a regular ideal of* R *that satisfies the following*

(1) *If* I *is a regular ideal of* R, *then* $H \otimes_R I$ *is the completion of* I, *and* $H \otimes_R I \cong HI$. HI *is a regular ideal of* H; $\lambda^{-1}(HI) = I$: $H/HI \cong R/I$; *and* $(HI)/\lambda(I)$ *is torsion-free and divisible. Furthermore,* $\mathrm{Tor}_1^R(H,I) = 0$.

(2) *If* J *is an ideal of* H *such that* $\lambda^{-1}(J)$ *is a regular ideal of* R *then* $J = H(\lambda^{-1}(J))$ *and* $H/J \cong R/(\lambda^{-1}(J))$.

(3) *If* I_1 *and* I_2 *are regular ideals of* R, *then* $H(I_1 \cap I_2) = HI_1 \cap HI_2$ *and* $H(I_1 + I_2) = HI_1 + HI_2$.

(4) *If* R *is a quasi-local ring with maximal ideal* M, *then* H *is a quasi-local ring with maximal ideal* HM.

Proof. (1) We have a map $\varphi_1 : H \otimes_R I \to HI$ defined by $\varphi_1(f \otimes x) = fx$ for all $f \in H$ and $x \in I$; and we have a map $\varphi_2 : H \otimes_R R \to H$ defined in a similar fashion. We also have the canonical map $\varphi_3 : H \otimes_R R/I \to H/HI$ defined by $\varphi_3(f \otimes (r+I)) = fr + HI$. Thus we have a commutative diagram:

$$\begin{array}{ccccccccc} 0 & \to & H \otimes_R I & \to & H \otimes_R R & \to & H \otimes_R R/I & \to & 0 \\ & & \varphi_1 \downarrow & & \varphi_2 \downarrow & & \varphi_3 \downarrow & & \\ 0 & \to & HI & \longrightarrow & H & \longrightarrow & H/HI & \longrightarrow & 0 \end{array}$$

The top row is exact by Corollary 2.5, since $\mathrm{Tor}_1^R(H,R/I) = 0$. It is clear that the vertical maps are epimorphisms and that the middle vertical map is an isomorphism. It follows from diagram chasing that the end vertical maps are isomorphisms also. Therefore, $H \otimes_R I \cong HI$; and by Theorem 2.7, H/HI is isomorphic to R/I.

Now $\lambda(I) \subset HI$, and since H is a torsion-free R-module, $\lambda(I)$ contains a regular element of H. Therefore, HI is a regular ideal of H. Because H/HI is isomorphic to R/I, we see that H/HI is a torsion R-module of bounded order, and hence is a cotorsion R-module. H is also a cotorsion R-module, and thus HI is a cotorsion R-module by Theorem 1.5.

Let $I' = \lambda^{-1}(HI)$; then $I \subset I'$, and hence I' is a regular ideal of R. Thus by what has already been proved we have $H/HI' \cong R/I'$ and $H/HI \cong R/I$. But clearly $HI = HI'$, and thus $R/I' \cong R/I$. It follows from this that $I = I' = \lambda^{-1}(HI)$.

Because $\lambda^{-1}(HI) = I$, we have $HI \cap \lambda(R) = \lambda(I)$. By Corollary 2.5, $H/\lambda(R)$ is a divisible R-module, and thus $H = HI + \lambda(R)$. Therefore, $H/\lambda(R) = (HI+\lambda(R))/\lambda(R) \cong HI/(HI \cap \lambda(R)) = HI/\lambda(I)$, and hence $HI/\lambda(I)$ is a torsion-free and divisible R-module.

If $\mathcal{D}$ is the divisible submodule of R, then $\mathcal{D} \subset I$ because I is a regular ideal of R. Therefore $\lambda(I) \cong I/\mathcal{D} = \overline{I}$. If $\tilde{I}$ is the completion of I in the R-topology and $\eta : I \to \tilde{I}$ is the map of Theorem 2.1, then $\operatorname{Ker} \eta = \mathcal{D}$, and hence $\eta(I) \cong I/\mathcal{D} = \overline{I}$. By Theorem 2.1, $\tilde{I}/\overline{I}$ is torsion-free and divisible and $\tilde{I}$ is a cotorsion R-module. We have proved that $(HI)/\overline{I}$ is also torsion-free and divisible and that HI is a cotorsion R-module. Hence it follows from Theorem 1.10 that $\tilde{I}$ is isomorphic to HI.

By Theorem 1.1 we have $\operatorname{Tor}_1^R(H,K) = 0$; and by Theorem 2.7 we have $H \otimes_R K \cong K$. Thus we have an exact sequence:

$$0 \to H \to H \otimes_R Q \to K \to 0.$$

Since $\operatorname{Tor}_n^R(K,I) = 0$ for all $n > 0$ by Theorem 1.1, we obtain from this sequence an isomorphism: $\operatorname{Tor}_1^R(H,I) \cong \operatorname{Tor}_1^R(H \otimes_R Q,I)$. By [3, Ch. VI, Prop. 4.1.2] we have $\operatorname{Tor}_1^R(H \otimes_R Q,I) \cong \operatorname{Tor}_1^Q(H \otimes_R Q,Q \otimes_R I)$. But since I is a regular ideal, we have $Q \otimes_R I \cong QI = Q$, and thus $\operatorname{Tor}_1^Q(H \otimes_R Q,Q \otimes_R I) = 0$. It follows that $\operatorname{Tor}_1^R(H,I) = 0$.

(2) Let J be a regular ideal of H such that $I = \lambda^{-1}(J)$ is a regular ideal of R. Since $H/\lambda(R)$ is a divisible R-module, we have $H = HI + \lambda(R)$. Consequently, $J = HI + \lambda(I) = HI$. Therefore, by (1) we see that $H/J = H/HI \cong R/\lambda^{-1}(J)$.

(3) Let I_1 and I_2 be regular ideals of R. Then $I_1 \cap I_2$ and $I_1 + I_2$ are also regular ideals of R, and we have an exact sequence:

$$0 \to I_1 \cap I_2 \to I_1 \oplus I_2 \to I_1 + I_2 \to 0.$$

By (1) we have $\mathrm{Tor}_1^R(H, I_1 + I_2) = 0$. Hence we obtain a commutative diagram with exact rows:

$$\begin{array}{ccccccccc} 0 \to & H \otimes_R (I_1 \cap I_2) & \to & (H \otimes_R I_1) \oplus (H \otimes_R I_2) & \to & H \otimes_R (I_1 + I_2) & \to 0 \\ & \downarrow & & \downarrow & & \downarrow & \\ 0 \longrightarrow & HI_1 \cap HI_2 & \longrightarrow & HI_1 \oplus HI_2 & \longrightarrow & HI_1 + HI_2 & \longrightarrow 0 \end{array}$$

The middle vertical map is an isomorphism by (1). Since it is obvious that $HI_1 + HI_2 = H(I_1 + I_2)$, the right hand vertical map is also an isomorphism by (1). Thus the left hand vertical map is an isomorphism. Because the image of this map is $H(I_1 \cap I_2)$, we have $H(I_1 \cap I_2) = HI_1 \cap HI_2$.

(4) Suppose that R is a quasi-local ring with maximal ideal M. By (1) HM is a maximal ideal of H. Let $\tilde{x} = \langle x_r + rR \rangle$ be an element of HM, where $x_r \in R$. It is easy to see that if some x_s is a unit in R, then every x_r is a unit in R. But then $\tilde{x}$ would be a unit in H, contradicting the fact that HM is a proper ideal of H. Thus every x_r is in M. But then every element $1 - x_r$ is a unit in R, and hence $1 - \tilde{x}$ is a unit in H. This shows that HM is the only maximal ideal of H.

<u>Theorem 2.9. Let S be a ring extension of R in Q, S $\neq$ Q, and let H(S) denote the completion of S in the S-topology. Then H(S) is a torsion-free R-module and we have an exact sequence:</u>

$$0 \to \mathrm{Hom}_R(K, S/R) \to H \xrightarrow{\varphi} H(S) \xrightarrow{\delta} \mathrm{Ext}_R^1(K, S/R)$$

<u>where φ is a ring homomorphism.</u>

(1) <u>If S/R has bounded order the sequence becomes:</u>

$$0 \to H \xrightarrow{\varphi} H(S) \to S/R \to 0.$$

(2) <u>If</u> $hd_R Q = 1$, <u>then</u> δ <u>is onto</u>.

(3) <u>If</u> $hd_R Q = 1$ <u>and</u> S/R <u>is</u> h-<u>divisible</u>, <u>the sequence becomes</u>:

$$0 \to \operatorname{Hom}_R(K,S/R) \to H \xrightarrow{\varphi} H(S) \to 0.$$

<u>Proof</u>. By Theorem 2.2, $H(S) \cong \operatorname{Hom}_S(Q/S,Q/S)$. We shall show first that $\operatorname{Hom}_S(Q/S,Q/S) = \operatorname{Hom}_R(Q/S,Q/S)$. Let $g \in \operatorname{Hom}_R(Q/S,Q/S)$ and $s \in S$. Then $s = a/b$ where $a, b \in R$ and b is regular in R. Define an R-homomorphism $\xi : Q/S \to Q/S$ by $\xi(x) = sg(x) - g(sx)$ for all $x \in Q/S$. Then $b\xi = 0$, and hence $\xi = 0$ since Q/S is a divisible R-module. Thus $g \in \operatorname{Hom}_S(Q/S,Q/S)$, proving the desired equality. $H(S)$ is thus a torsion-free R-module, since Q/S is a divisible R-module.

By Theorem 2.7, Q/S is an H-module and $\operatorname{Hom}_R(Q/S,Q/S) = \operatorname{Hom}_H(Q/S,Q/S)$. Thus we have a ring homomorphism $\varphi : H \to H(S)$ defined by $\varphi(f)(x) = fx$ for all $f \in H$ and $x \in Q/S$. An argument similar to that given in Theorem 2.7 shows easily that $\ker \varphi = \operatorname{Hom}_R(K,S/R)$.

Let π be the canonical R-homomorphism of K onto Q/S with $\ker \pi = S/R$. Then from the exact sequence $0 \to S/R \to K \xrightarrow{\pi} Q/S \to 0$ we obtain a diagram:

$$\begin{array}{ccccccc} \operatorname{Hom}_R(K,K) & \xrightarrow{\eta} & \operatorname{Hom}_R(K,Q/S) & \xrightarrow{\nu} & \operatorname{Ext}^1_R(K,S/R) & \to & \operatorname{Ext}^1_R(K,K) \\ & \varphi \searrow & & \nearrow \lambda & & & \\ & & \operatorname{Hom}_R(Q/S,Q/S) & & & & \end{array}$$

where η and λ are induced by π, and the top row is exact. It is easily verified that the diagram commutes. We let $\delta = \nu\lambda$; then, because λ is obviously a monomorphism, it is easily checked that $\operatorname{Im} \varphi = \ker \delta$. This establishes the first exact sequence of the theorem.

In fact, λ is an isomorphism. For let $f \in \mathrm{Hom}_R(K,Q/S)$. Then it is easily verified that if $q \in Q$, then there exists an element $s_q \in S$ such that $f(q + R) = s_q q + S$. Thus $f(S/R) = 0$, and hence f induces an R-homomorphism $g : Q/S \to Q/S$. But then $\lambda(g) = f$, and hence λ is onto. Since we have already observed that λ is a monomorphism, we see that λ is an isomorphism. This proves that $\mathrm{Im}\ \nu = \mathrm{Im}\ \delta$. If $\mathrm{hd}_R Q = 1$, then $\mathrm{Ext}_R^1(K,K) = 0$ by Theorem 1.7. The diagram then shows that ν is onto, and hence δ is onto. If in addition S/R is h-divisible, then $\mathrm{Ext}_R^1(K,S/R) = 0$ by Theorem 1.7, and hence (3) is established.

Suppose now that S/R has bounded order. Then, of course, $\mathrm{Hom}_R(K,S/R) = 0$. Since $\mathrm{Ext}_R^1(Q,S/R) = 0$, we see by Theorem 1.1 (3) that $S/R \cong \mathrm{Ext}_R^1(K,S/R)$. We also see by Theorem 1.1 (3) that $\mathrm{Ext}_R^1(K,K) \cong \mathrm{Ext}_R^1(Q,K)$ and hence $\mathrm{Ext}_R^1(K,K)$ is torsion-free. The diagram then shows that ν is onto, and hence δ is onto in this case also. Thus we have proved that the sequence

$$0 \to H \xrightarrow{\varphi} H(S) \to S/R \to 0$$

is exact.

Theorem 2.10. H is a subring of $H \otimes_R Q$; and if $\mathcal{D}$ is the divisible submodule of R, then $Q/\mathcal{D} \subset H \otimes_R Q$ and there is a one-to-one correspondence between the set of rings B between R and Q and the set of rings Ω between H and $H \otimes_R Q$ given by $\Omega = H \otimes_R B$ and $B/\mathcal{D} = \Omega \cap Q/\mathcal{D}$. Furthermore:

(1) B/R and $(H \otimes_R B)/H$ are isomorphic torsion modules.

(2) Q/B and $(H \otimes_R Q)/(H \otimes_R B)$ are isomorphic torsion modules.

(3) If $\overline{B} = B/\mathcal{D}$, then $\overline{B} \subset H \otimes_R B$, and $H \otimes_R B = H\overline{B} = H + \overline{B}$.

(4) $(H \otimes_R B)/\overline{B}$ is a torsion-free and divisible R-module.

Proof. Since Q/B and B/R are torsion R-modules, we have $\mathrm{Tor}_1^R(H,Q/B) = 0$ and $\mathrm{Tor}_1^R(H,B/R) = 0$ by Corollary 2.5 (3); and we have $H \otimes_R Q/B \cong Q/B$ and $H \otimes_R B/R \cong B/R$ by Theorem 2.7 (1). Thus we have exact sequences:

$$0 \to H \to H \otimes_R B \to B/R \to 0 \text{ and}$$

$$0 \to H \otimes_R B \to H \otimes_R Q \to Q/B \to 0.$$

If we let $B = Q$, then we see that $H \subset H \otimes_R Q$. In general we see that $H \otimes_R B$ is a ring between H and $H \otimes_R Q$, and we have proved (1) and (2) of the theorem.

Now if V is any Q-module, then $\mathrm{Tor}_1^R(B,V) = 0$. For by [3, Ch. VI, Prop. 4.1.2] we have $\mathrm{Tor}_1^R(B,V) \cong \mathrm{Tor}_1^Q(Q \otimes_R B,V)$; and since $Q \otimes_R B \cong Q$, it follows that $\mathrm{Tor}_1^R(B,V) = 0$. In particular, we have $\mathrm{Tor}_1^R(B,\mathrm{Ext}_R^1(Q,R)) = 0$. Therefore if we let $\overline{R} = R/\mathcal{O}$ and tensor the exact sequence of Corollary 2.5(2) with B we obtain an exact sequence

$$0 \to \overline{R} \otimes_R B \to H \otimes_R B \to \mathrm{Ext}_R^1(Q,R) \otimes_R B \to 0.$$

Now $\mathcal{O}$ is an ideal of Q, and hence $\mathcal{O}B = \mathcal{O}$. Therefore, if $\overline{B} = B/\mathcal{O}$, then $\overline{B} = B/\mathcal{O} \cong \overline{R} \otimes_R B$. Thus we have proved that $\overline{B} \subset H \otimes_R B$ and that $(H \otimes_R B)/\overline{B}$ is a torsion-free and divisible R-module.

The previous exact sequence shows that if $x \in B$ and $\overline{x} = x + \mathcal{O}$, then we may identify $\overline{x}$ with $1 \otimes x$ in $H \otimes_R Q$. Furthermore, since $H \subset H \otimes_R Q$, we may identify an element $h \in H$ with its image $h \otimes 1$ in $H \otimes_R Q$. If $\lambda \in H \otimes_R B$, then $\lambda = \sum_{i=1}^{n} h_i \otimes x_i$, where $h_i \in H$ and $x_i \in B$. Hence $\lambda = \sum_{i=1}^{n} h_i\overline{x}_i \in H\overline{B}$, showing that $H \otimes_R B = H\overline{B}$. Suppose that $h \in H$ and $x \in B$. Then $x = a/b$, where $a \in R$ and b is a regular element of R. Because $H/\overline{R}$ is a divisible R-module by Corollary 2.5, there exist elements $g \in H$ and $r \in R$ such that $h = bg + \overline{r}$ (where $\overline{r} = r + \mathcal{O} \in \overline{R}$). Thus $h \otimes x = (bg + \overline{r}) \otimes x = bg \otimes x + (\overline{r} \otimes x)$ $= (g \otimes a) + (1 \otimes rx) = (ag \otimes 1) + (1 \otimes rx) \in H + \overline{B}$. Thus we have shown that $H \otimes_R B = H + \overline{B}$.

We identify $\overline{Q} = Q/\mathcal{O}$ with a subring of $H \otimes_R Q$ as we have done with $\overline{B}$. Then of course we have $\overline{B} \subset (H \otimes_R B) \cap \overline{Q}$. On the other hand, suppose that $\lambda \in (H \otimes_R B) \cap \overline{Q}$. Then there exists a regular element b in R such that $b\lambda \in \overline{B}$. However, we have proved that $(H \otimes_R B)/\overline{B}$ is torsion-free, and thus $\lambda \in \overline{B}$. This proves that $(H \otimes_R B) \cap \overline{Q} = \overline{B}$.

Let Ω be a ring between H and $H \otimes_R Q$, and let $\overline{B} = \Omega \cap \overline{Q}$. Then $\overline{B} = B/\mathcal{D}$, where B is a ring between R and Q, and to complete the proof of the theorem it is sufficient to show that $\Omega = H + \overline{B}$. Let $\lambda \in \Omega$; then because $(H \otimes_R Q)/H \cong K$ is a torsion R-module, there exists a regular element $b \in R$ such that $b\lambda = h \in H$. But $H/\overline{R}$ is a divisible R-module, and hence there exist elements $g \in H$ and $r \in R$ such that $h = bg + \overline{r}$. Thus $b(\lambda - g) = \overline{r} \in \overline{R}$, and hence $\lambda - g = \overline{r/b} \in \Omega \cap \overline{Q} = \overline{B}$. Therefore, $\lambda = g + \overline{r/b} \in H + \overline{B}$, proving that $\Omega = H + \overline{B}$. Since $H + \overline{B} = H \otimes_R B$, the proof of the theorem is complete.

Theorem 2.11. *Let $\mathcal{D}$ be the divisible submodule of R and let $\overline{R} = R/\mathcal{D}$, $\overline{Q} = Q/\mathcal{D}$, and $\overline{K} = \overline{Q}/\overline{R}$. Assume that every regular element of $\overline{Q}$ is invertible in $\overline{Q}$. Then*

(1) $\overline{Q}$ is the full ring of quotients of $\overline{R}$; $\overline{K} \cong K$; and $\overline{H}$, the completion of $\overline{R}$, is isomorphic to H, the completion of R.

(2) The categories of torsion h-divisible R-modules and torsion-free, cotorsion R-modules are equal to the same categories of $\overline{R}$-modules.

(3) I is a regular ideal of R, if and only if $\overline{I} = I/\mathcal{D}$ is a regular ideal of $\overline{R}$ and then the completions of I and $\overline{I}$ in the R and $\overline{R}$ topologies, respectively, are isomorphic.

Proof. Now $\mathcal{D}$ is torsion-free and divisible, and thus $\mathcal{D}$ is an ideal of Q. Therefore $\overline{Q}$ is a ring; and by assumption, $\overline{Q}$ is equal to its own full ring of quotients.

Let b be a regular element of R, and $\overline{b} = b + \mathcal{D}$, the image of b in $\overline{R}$. If $a \in R$ and $ba \in \mathcal{D}$, then $ba = bd$ for some element $d \in \mathcal{D}$, since $\mathcal{D}$ is a divisible R-module. But then $b(a-d) = 0$, and hence $a = d$ because R is torsion-free. This shows that $\overline{b}$ is a regular element of $\overline{R}$. If we let $\overline{\mathcal{S}}$ be the set of all elements $\overline{b}$ in $\overline{R}$ such that b is a regular element of R, then we have $\overline{Q} \cong \overline{R}_{\overline{\mathcal{S}}}$. Since $\overline{\mathcal{S}}$

is a subset of the regular elements of $\overline{R}$, it follows that $\overline{Q}$ is contained in the full ring of quotients of $\overline{R}$. However, $\overline{Q}$ is equal to its own full ring of quotients by assumption, and thus $\overline{Q}$ is the full ring of quotients of $\overline{R}$.

Since $Q\mathscr{P} = \mathscr{P} \subset R$, we have $\mathscr{P}K = (Q\mathscr{P} + R)/R = 0$. Hence K is an $\overline{R}$-module, and of course $K \cong \overline{K}$. Furthermore, $\mathrm{Hom}_R(K,K) = \mathrm{Hom}_{\overline{R}}(\overline{K},\overline{K})$, and thus R and $\overline{R}$ have isomorphic completions. Every torsion h-divisible R-module is of the form $K \otimes_R A$ for some R-module A by Corollary 1.2, and clearly $K \otimes_R A \cong \overline{K} \otimes_R A \cong \overline{K} \otimes_{\overline{R}} (\overline{R} \otimes_R A)$ is a torsion h-divisible $\overline{R}$-module. On the other hand if $\overline{A}$ is an $\overline{R}$-module, then $\overline{R} \otimes_R \overline{A} \cong \overline{A}$, and hence $K \otimes_R \overline{A} \cong \overline{K} \otimes_{\overline{R}} \overline{A}$ is a torsion h-divisible R-module. In similar fashion, using Corollary 2.3, we can prove that an R-module is a torsion-free cotorsion R-module if and only if it is a torsion-free cotorsion $\overline{R}$-module.

Let I be a regular ideal of R and let $\overline{I} = I/\mathscr{P}$. Then there is a regular element b of R that is contained in I. In the first paragraph of the proof we showed that $\overline{b} = b + \mathscr{P}$ is a regular element of $\overline{R}$, and hence $\overline{I}$ is a regular ideal of $\overline{R}$. Conversely, let $\overline{I}$ be a regular ideal of $\overline{R}$. Then there is an ideal I of R such that $\overline{I} = I/\mathscr{P}$. Let $\overline{c}$ be a regular element of $\overline{I}$; then $\overline{c} = c + \mathscr{P}$, where $c \in I$. Since $\overline{Q}$ is the full ring of quotients of $\overline{R}$, there is an element $\overline{q}$ of $\overline{Q}$ such that $\overline{cq} = 1$. Now $\overline{q} = q + \mathscr{P}$, where $q \in Q$ and hence $q = a/b$ where $a \in R$ and b is a regular element of R. Thus $ca/b - 1 \in \mathscr{P}$, and hence $ca - b \in \mathscr{P}$. Therefore b is an element of I, showing that I is a regular ideal of R.

If I is a regular ideal of R we have $K \otimes_R I \cong \overline{K} \otimes_{\overline{R}} \overline{I}$, and hence $\mathrm{Hom}_R(K, K \otimes_R I) \cong \mathrm{Hom}_{\overline{R}}(\overline{K}, \overline{K} \otimes_{\overline{R}} \overline{I})$. Thus by Theorem 2.2 the completion of I in the R-topology is isomorphic to the completion of $\overline{I}$ in the $\overline{R}$-topology.

<u>Theorem 2.12</u>. H <u>is complete in the</u> H-<u>topology</u>. <u>Furthermore</u>, <u>if</u>

<u>an</u> R-<u>module</u> <u>is</u> <u>a</u> <u>cotorsion</u> (<u>or</u> <u>torsion</u>) R-<u>module</u> <u>then</u> <u>it</u> <u>is</u> <u>also</u> <u>a</u> <u>cotorsion</u> (<u>or</u> <u>torsion</u>) H-<u>module</u>.

<u>Proof</u>. By Theorem 2.7 a cotorsion (or torsion) R-module is an H-module. Let $\lambda : R \to H$ be the ring homomorphism of Corollary 2.5. If b is a regular element of R, then $\lambda(b)$ is a regular element of H because H is a torsion-free R-module. Thus a torsion R-module is also a torsion H-module. By Corollary 2.3 if we prove that H is a cotorsion H-module, then we will have shown that H is complete in the H-topology. Hence to finish the proof of the theorem, it is sufficient to prove that if C is a cotorsion R-module, then C is a cotorsion H-module.

Let V be the full ring of quotients of H; then V is a torsion-free and divisible R-module, and hence V is a Q-module. Thus $\mathrm{Hom}_H(V,C) \subset \mathrm{Hom}_R(V,C) = 0$. Therefore it is sufficient to prove that $\mathrm{Ext}^1_H(V,C) = 0$. Let us consider an exact sequence of the form:

$$0 \to C \xrightarrow{f} A \xrightarrow{g} V \to 0$$

where A is an H-module, and f and g are H-homomorphisms. Since $\mathrm{Ext}^1_R(V,C) = 0$, there is an R-homomorphism $s : A \to C$ such that $s \circ f$ is the identity on C. But $\mathrm{Hom}_R(A,C) = \mathrm{Hom}_H(A,C)$ by Theorem 2.7, and hence s is an H-homomorphism. Therefore, the given exact sequence splits over H, showing that $\mathrm{Ext}^1_H(V,C) = 0$.

CHAPTER III

COMPATIBLE EXTENSIONS

Theorem 3.1. *Let A be a commutative ring extension of R that is torsion-free as an R-module. Then the following statements are equivalent:*

(1) *If J is a regular ideal of A, then $J \cap R$ is a regular ideal of R.*

(2) *If x is a regular element of A, then there is an element $y \in A$ such that $xy = b$ is a regular element of R.*

(3) *$A \otimes_R Q$ is the full ring of quotients of A.*

Proof. (1) ==> (2). If x is a regular element of A, then Ax is a regular ideal of A and hence $Ax \cap R$ is a regular ideal of R by (1). Thus there is an element $y \in A$ such that $yx = b$ is a regular element of R.

(2) ==> (1). Let J be a regular ideal of A, and x a regular element of J. By (2) there is a regular element of R in $Ax \cap R$, and hence $J \cap R$ is a regular ideal of R.

(2) ==> (3). If λ is an element of the full ring of quotients of A, then $\lambda = z/x$, where $z \in A$ and x is a regular element of A. By (2) there is an element $y \in A$ such that $xy = b$ is a regular element of R. Since A is a torsion-free R-module, both b and y are regular in A. Hence $\lambda = zy/xy = zy/b$. Since $A \otimes_R Q = A_Q$ is contained in the full ring of quotients of A, this expression for λ shows that $A \otimes_R Q$ is equal to the full ring of quotients of A.

(3) ==> (2). Let x be a regular element of A. Then by (3), there is an element $y \in A$ and a regular element $b \in R$ such that $1/x = y/b$. Hence $xy = b$, proving (2).

Theorem 3.2. *Let A be a commutative ring extension of R that is torsion-free as an R-module, and assume that if J is a regular ideal of A, then $J \cap R$ is a regular ideal of R. Then the R-topology and* *the A-topology on any A-module coincide. Furthermore, an A-module has one of the following properties as an A-module if and only if it has the same property as an R-module:*

(1) *torsion*, (2) h-*reduced*, (3) *cotorsion*, (4) *complete*, (5) *divisible*, (6) h-*divisible*.

Proof. It is obvious from Theorem 3.1 that the R-topology and the A-topology on an A-module coincide and that (1), (4), and (5) are true.

Let B be an A-module. By [3, Ch. VI, Prop. 4.1.3], we have $\mathrm{Ext}_R^n(Q,B) \cong \mathrm{Ext}_A^n(A \otimes_R Q,B)$ for all $n \geq 0$. Since $A \otimes_R Q$ is the ring of quotients of A by Theorem 3.1, (2) and (3) are seen to be true. We have a commutative diagram:

$$\begin{array}{ccc} \mathrm{Hom}_A(A \otimes_R Q,B) & \xrightarrow{\varphi} & \mathrm{Hom}_R(Q,B) \\ & \lambda_A \searrow \quad \swarrow \lambda_R & \\ & B & \end{array}$$

where φ is the canonical duality isomorphism, λ_R is defined by $\lambda_R(f) = f(1)$ for $f \in \mathrm{Hom}_R(Q,B)$, and λ_A is defined similarly. By Theorem 1.1, B is h-divisible over R (or over A) if and only if λ_R (resp. λ_A) is an epimorphism. Thus we see that (6) is true.

Theorem 3.3. Suppose that A is an integral domain containing R and that the quotient field of A is algebraic over Q. Then the quotient field of A is isomorphic to $A \otimes_R Q$, and thus all of the statements of Theorem 3.2 are true for R and A.

Proof. If x is a nonzero element of A, then the minimum polynomial for x over Q has a nonzero constant term. From this it follows readily that if J is a nonzero ideal of A, then $J \cap R \neq 0$. Theorem 3.3 is thus a consequence of Theorems 3.1 and 3.2.

Theorem 3.4. Let A be a commutative ring extension of R, that is torsion-free as an R-module and assume that if J is any regular ideal of A, *then* J ∩ R *is a regular ideal of* R. *Then the following statements are equivalent:*

(1) A = J + R *for every regular ideal* J *of* A.

(2) A/R *is a divisible* R-*module*.

(3) A/J ≅ R/(J ∩ R) *for every regular ideal* J *of* A.

If any of these statements are true, then J = A(J ∩ R) *for every regular ideal* J *of* A.

Proof. (1) ⟹ (2). If b is a regular element of R, then Ab is a regular ideal of A, and hence A = Ab + R. Thus b(A/R) = A/R, showing that A/R is a divisible R-module.

(2) ⟹ (1). If J is a regular ideal of A, choose a regular element b ∈ J ∩ R. Then b(A/R) = A/R, and hence A = Ab + R ⊂ J + R ⊂ A.

(1) ⟹ (3). Let J be a regular ideal of A; then A/J = (J+R)/J ≅ R/(J ∩ R).

(3) ⟹ (1). Let J be a regular ideal of A. Since A/J is a cyclic R-module by assumption, there is an element x ∈ A such that A = Rx + J. Thus 1 = rx + j, where r ∈ R and j ∈ J. Therefore, we have A = Ar + J. We also have Ar = Rxr + Jr, and hence A = Rxr + J. But Rxr + J = R + J, and thus A = R + J.

Assume that any one of these equivalent statements is true, and let J be a regular ideal of A. If we let I = A(J ∩ R), then I is a regular ideal of A and I ⊂ J. Thus by (1) we have A = I + R. Hence J = I + (R ∩ J) ⊂ I, and we see that J = I.

Definitions: Let A be a commutative ring extension of R. We shall say that A is a *compatible extension of R* if it satisfies the following conditions:

(1) A is torsion-free as an R-module.

(2) A is not equal to its full ring of quotients.

(3) If J is a regular ideal of A, then J ∩ R is a regular ideal of R.

(4) $A = J + R$ for every regular ideal J of A.

We shall say that R is a closed domain if R is an integral domain, and if $J \cap R \neq 0$ for every nonzero ideal J of H. Then, of course, H is also an integral domain. Since H is a torsion-free R-module, and H/R is a divisible R-module, we see by Theorem 3.4 that in this case H is a compatible extension of R. In particular, all of the statements of Theorem 3.2 are then true about R and H. By Theorem 3.1, the quotient field of H is equal to $H \otimes_R Q$.

If R is a closed domain, then every proper R-submodule of K is h-reduced. For if D is a proper, nonzero, h-divisible R-submodule of K, then $J = \mathrm{Hom}_R(K,D)$ is a nonzero ideal of H by Corollary 1.3. However, $J \cap R = 0$, since every nonzero element of R acts as an epimorphism on K. This contradiction proves that every proper submodule of K is h-reduced.

We have proved [9, Theorem 4.5] that if R is a quasi-local domain, then the converse is true; namely, if every proper R-submodule of K is h-reduced, then R is a closed domain.

Corollary 3.5. *Let A be a proper R-submodule of Q containing R (A not assumed to be a ring). Then A is a compatible ring extension of R if and only if A/R is a divisible R-module.*

Proof. Suppose that A/R is a divisible R-module. Let x be a nonzero element of A; then $x = a/b$, where $a \in R$ and b is a regular element of R. Then $A = Ab + R$, and hence $Ax = Aa + Rx$. Since Aa and Rx are contained in A, we see that $Ax \subset A$. Thus A is a ring. Clearly every regular ideal of A has a regular intersection with R, and A is a torsion-free R-module. Therefore, A is a compatible ring extension of R by Theorem 3.4. The converse statement also follows from Theorem 3.4.

Theorem 3.6. *Let A be a compatible ring extension of R. Then:*

(1) $A \otimes_R Q$ *is the full ring of quotients of* A, *and all of the statements of Theorem* 3.2 *are true for* R *and* A.

(2) *If* J *is a regular ideal of* A, *then* $A = J + R$ *and* $A/J \cong R/(J \cap R)$. *Furthermore* $J = A(J \cap R)$ *and* $J \cap R$ *is a regular ideal of* R.

(3) *A finitely generated torsion* A-*module is also a finitely generated torsion* R-*module*.

(4) *If every regular ideal of* R *is finitely generated over* R, *then every regular ideal of* A *is finitely generated over* A.

(5) N *is a regular maximal* (*resp. prime*) *ideal of* A *if and only if* $N \cap R$ *is a regular maximal* (*resp. prime*) *ideal of* R.

(6) *If* R *is a quasi-local ring with maximal ideal* M, *then* AM *is a regular maximal ideal of* A *and* $A/AM \cong R/M$. AM *contains every regular ideal of* A. *If* A *is a reduced* R-*module, then* AM *is the only maximal ideal of* A *and* A *is a quasi-local ring*.

(7) *Let* S *be an* R-*submodule of* Q *containing* R. *Then in* $A \otimes_R Q$ *we have* $AS = A + S$; *and* AS/S *is a divisible* R-*module*. *If* S *is a proper subring of* Q *containing* R, *then* AS/S *is a divisible* S-*module*. *Thus if* $AS \neq A \otimes_R Q$, *then* AS *is a compatible ring extension of* S. *If* $A \subset S$, *then* S *is also an* A-*module*.

(8) *If* R *is an integral domain, if* $A \subset Q$, *and if* V *is a valuation subring of* Q *containing* R, *then either* $A + V = Q$ *or* $A \subset V$.

Proof. (1), (2), (4) and (5) are immediate consequences of Theorems 3.1 and 3.4. As for (3), a cyclic torsion A-module is a cyclic torsion R-module by Theorem 3.4, and the general statement follows easily by induction on the number of generators.

(6). Assume that R is quasi-local and let M be the maximal ideal of R. Let N be any regular maximal ideal of A. By (5), $N \cap R$ is a regular maximal ideal of R and hence $N \cap R = M$. By (2) we have $AM = A(N \cap R) = N$, and thus AM is the only regular maximal ideal of A and contains every regular ideal of A. By (2) we have $A/AM \cong R/M$.

If A is not a quasi-local ring, then there is a proper ideal I of A such that $I \not\subset AM$. Choose an element $y \in I$ such that $y \notin AM$. Since I is not a regular ideal of A by the preceding paragraph, y is a zero divisor in A, and hence there exists a nonzero element $z \in A$ such that $yz = 0$. Let b be any regular element of R. Since $Ab + Ay$ is a regular ideal of A and is not contained in AM, we have $Ab + Ay = A$. Thus $Az = bAz$. Hence Az is a divisible submodule of A. If we assume A to be reduced, this contradiction shows that A is a quasi-local ring with maximal ideal AM.

(7). Let S be an R-submodule of Q containing R. Let $q \in S$; then $q = a/b$ where $a \in R$ and b is regular in R. Because A/R is a divisible R-module, we have $Aq = Aa + Rq$. Thus $Aq \subset A + S$, and hence $AS \subset A + S$. Since $1 \in A \cap S$, we have the reverse inequality. We then have $AS/S = (A + S)/S \cong A/(S \cap A)$, and hence AS/S is a homomorphic image of A/R. Therefore, AS/S is a divisible R-module.

Suppose that S is a proper subring of Q containing R. It follows from Theorem 3.2 that since AS/S is a divisible R-module it is also a divisible S-module. It is obvious that the intersection with S of a regular ideal of AS is a regular ideal of S. Therefore, if $AS \neq A \otimes_R Q$, then AS is a compatible ring extension of S by Theorem 3.4.

If $A \subset S$, then $AS = A + S = S$, and hence S is an A-module.

(8). Assume that R is an integral domain, that $A \subset Q$, and that V is a valuation subring of Q containing R. If $A + V \neq Q$, then $AV = A + V$ is a proper V-submodule of Q. Hence $(AV)/V$ is a torsion V-module of bounded order. By (6) AV/V is a divisible V-module, and thus $AV = V$. Hence we have $A \subset V$.

CHAPTER IV

LOCALIZATIONS

Definitions: A commutative ring is said to have Krull dimension $n < \infty$ if every chain of prime ideals in R has at most $n + 1$ terms, and there is at least one chain with exactly $n + 1$ terms. We will say that a ring R is a 1-dimensional Cohen-Macaulay ring if it is a Noetherian ring of Krull dimensional 1, and if every maximal ideal of R contains a regular element. A Noetherian domain of Krull dimension 1 is a 1-dimensional Cohen-Macaulay ring. Such a domain differs from a Dedekind ring only in the fact that it may not be integrally closed.

Let R be a Noetherian ring of Krull dimension 1, and let T be a torsion R-module. If M_α is a maximal ideal of R, define the M_α-primary component of T by:

$$T_\alpha = \{x \in T \mid x = 0, \text{ or } \mathrm{Ann}_R(x) \text{ is an } M_\alpha\text{-primary ideal}\}$$

where $\mathrm{Ann}_R(x) = \{r \in R \mid rx = 0\}$. Clearly T_α is a submodule of T.

Theorem 4.1. *Let R be a 1-dimensional Cohen-Macaulay ring and let T be a torsion R-module. Then*

$$T = \sum_\alpha \oplus T_\alpha$$

where M_α ranges over all of the maximal ideals of R. Furthermore, $T_\alpha \cong T_{M_\alpha}$; and hence T_α is an R_{M_α}-module.

Proof. Let x be a nonzero element of T, and let $I = \mathrm{Ann}_R(x)$. Take a primary decomposition of I; $I = J_1 \cap \ldots \cap J_n$ where J_i is M_i-primary. If $L_i = \cap J_k$, $(k \neq i)$, then $R = L_1 + \ldots + L_n$. Hence we can write $1 = r_1 + \ldots + r_n$, where $r_i \in L_i$. Thus we have $x = r_1x + \ldots + r_nx$. Since $J_ir_i \subset I$, we have $J_ir_ix = 0$ for $i = 1,\ldots,n$. Therefore, either $\mathrm{Ann}_R(r_ix)$ contains a power of M_i and hence is an M_i-primary ideal, or $r_ix = 0$. This proves that T is the sum of the T_α's.

We prove next that the sum is direct. Let $M_1,\ldots,M_n$ be distinct maximal ideals of R, and suppose that $x \in T_1 \cap (\sum_{j>1} T_j)$. Then there are positive integers $k_1,\ldots,k_n$ such that $M_1^{k_1}x = 0$ and $M_2^{k_2} \ldots M_n^{k_n}x = 0$. Since $R = M_1^{k_1} + (M_1^{k_2} \ldots M_n^{k_n})$, we see that $x = 0$. Hence we have shown that $T = \sum_\alpha \oplus T_\alpha$.

It is obvious that T_α is uniquely divisible by the elements of $R - M_\alpha$, and hence $(T_\alpha)_{M_\alpha} \cong T_\alpha$. On the other hand, if $\beta \neq \alpha$, we have $(T_\beta)_{M_\alpha} = 0$. Thus $T_{M_\alpha} \cong \sum_\beta \oplus (T_\beta)_{M_\alpha} = (T_\alpha)_{M_\alpha} \cong T_\alpha$.

Theorem 4.2. Let R be a 1-dimensional Cohen-Macaulay ring. Then $hd_R Q = 1$.

Proof. Case I. R is a local ring.

Let b be a regular element of the maximal ideal of R, and let $\mathcal{S} = \{b^n\}$. Then there is a one-to-one correspondence between the prime ideals of rank 0 in R and the prime ideals of $R_\mathcal{S}$. Hence $R_\mathcal{S}$ is a semi-local Noetherian ring and every prime ideal of $R_\mathcal{S}$ has rank 0 and is a maximal ideal of $R_\mathcal{S}$. Therefore, every regular element of $R_\mathcal{S}$ is a unit in $R_\mathcal{S}$, and hence $R_\mathcal{S}$ is equal to its own full ring of quotients. Since $R \subset R_\mathcal{S} \subset Q$, it follows that $R_\mathcal{S} = Q$.

Let F be a free R-module with a countable free basis $\{x_n\}$ and define an R-epimorphism $f : F \to Q$ by $f(x_n) = 1/b^n$. It is then easy to see that if P is the kernel of f, then P is a free R-module with basis $\{x_{n+1}-bx_n\}$. Hence we have $hd_R Q = 1$.

Case II. General case.

Let B be a torsion h-divisible R-module. In order to prove that $hd_R Q = 1$, it is sufficient by Theorem 1.7 to prove that $Ext_R^1(K,B) = 0$. Let us consider an exact sequence of the form:

$$(1) \qquad 0 \to B \to A \to K \to 0.$$

By Theorem 4.1 we have $B \cong \sum_\alpha \oplus B_{M_\alpha}$, $A \cong \sum_\alpha \oplus A_{M_\alpha}$ and $K \cong \sum_\alpha \oplus K_{M_\alpha}$, where M_α ranges over all of the maximal ideals of R. Thus exact sequence (1) is the direct sum of the exact sequences:

$$(2) \qquad 0 \to B_{M_\alpha} \to A_{M_\alpha} \to K_{M_\alpha} \to 0.$$

If every one of these sequences split, then (1) is also split. Hence because (2) is an R_{M_α}-sequence, it is sufficient to prove that $\mathrm{Ext}^1_{R_{M_\alpha}}(K_{M_\alpha}, B_{M_\alpha}) = 0$. If we prove that Q_{M_α} is the full ring of quotients of R_{M_α}, then we see that B_{M_α} is a torsion, h-divisible R_{M_α}-module, and the theorem will follow from Case I and Theorem 1.7.

Let M be a maximal ideal of R and let $C = \{x \in Q \mid sx = 0$ for some $s \in R - M\}$. Then C is an ideal of Q and we let $\overline{Q} = Q/C$. Since Q is a semi-local ring and every prime ideal of Q has rank 0 and is maximal in Q, the same is true of $\overline{Q}$. Thus $\overline{Q}$ is equal to its full ring of quotients. Let $I = R \cap C$; then $I \subset M$, and we let $\overline{R} = R/I$, and $\overline{M} = M/I$. We have $\overline{R} \subset \overline{Q}$, and $\overline{Q}$ is contained in the full ring of quotients of $\overline{R}$. But since $\overline{Q}$ is equal to its full ring of quotients, it follows that $\overline{Q}$ is the full ring of quotients of $\overline{R}$.

We observe that $R_M = \overline{R}_{\overline{M}}$ and $Q_M = \overline{Q}_{\overline{M}}$. Because no element of $\overline{R} - \overline{M}$ is a zero divisor in $\overline{R}$, we have $\overline{Q}_{\overline{M}} = \overline{Q}$, and hence $\overline{Q} = Q_M$ is the full ring of quotients of R_M.

<u>Theorem 4.3</u>. <u>Let</u> R <u>be</u> <u>a</u> <u>1-dimensional Cohen-Macaulay ring</u>. <u>Then an R-module is divisible if and only if it is h-divisible</u>.

<u>Proof</u>. Of course an h-divisible R-module is divisible, and hence we must prove the converse.

<u>Case I</u>: R is a local ring.

Let b be a regular element of the maximal ideal of R and let $\mathcal{S} = \{b^n\}$. As in the proof of Theorem 4.2 we then see that $Q = R_{\mathcal{S}}$. Let D be a divisible R-module, and let $x \in D$. Let $x = x_1$, and having found x_n, let x_{n+1} be an element of D such that $bx_{n+1} = x_n$. Define an R-homomorphism $f : Q \to D$ by $f(a/b^n) = ax_{n+1}$, where $a \in R$ and

$n \geq 0$. It is easily verified that f is a well-defined R-homomorphism and that $f(1) = x$. Thus D is a homomorphic image of a direct sum of copies of Q and thus D is an h-divisible R-module.

Case II: General Case.

Let D be a divisible R-module. Since $hd_R Q = 1$ by Theorem 4.2, we see that D/h(D) is h-reduced by Theorem 1.9. Thus we can assume that D is h-reduced, and we shall prove that D = 0. Let T = t(D), the torsion submodule of D; then by Theorem 4.1, $T \cong \sum_\alpha \oplus T_{M_\alpha}$, where M_α ranges over all of the maximal ideals of R. Since T_{M_α} is a torsion, divisible R_{M_α}-module, T_{M_α} is an h-divisible R_{M_α}-module by Case 1. But in the proof of Theorem 4.2 we saw that the full ring of quotients of R_{M_α} is a homomorphic image of Q, and thus T_{M_α} is an h-divisible R-module. Thus $T_{M_\alpha} = 0$ for all M_α, and hence D is torsion-free. But then D is h-divisible, and therefore D = 0.

Corollary 4.4. Let R be a 1-dimensional Cohen-Macaulay ring with no nilpotent elements other than zero. Then the torsion-submodule of every divisible R-module is a direct summand.

Proof. Let $P_1,\ldots,P_t$ be the prime ideals of rank 0 in R. Then $Q \cong R_{P_1} \oplus \ldots \oplus R_{P_t}$. Since R has no nonzero nilpotent elements, we have $0 = P_1 \cap \ldots \cap P_t$. From this it follows that R_{P_i} is a field for $i = 1,\ldots,t$, and thus Q is a semi-simple ring. The theorem is now an immediate consequence of Theorem 4.3 and Theorem 1.4.

Remarks. Corollary 4.4 and Theorem 4.1 show that in order to study divisible modules over a 1-dimensional Cohen-Macaulay ring with no nonzero nilpotent elements, it is sufficient to consider the case when R is a local ring. For if D is a divisible R-module, then $D = t(D) \oplus D/t(D)$. Now D/t(D) is a Q-module, and Q is a finite direct sum of fields. Hence D/t(D) is a direct sum of vector spaces over the fields that are direct summands of Q, and hence can be completely described by the dimensions of these vector spaces over these fields. For the torsion part, $t(D) = \sum_{\alpha} \oplus T_\alpha$ is a direct sum of divisible, torsion R_{M_α}-modules T_α, where M_α ranges over all of the maximal ideals M_α of R. Hence if we know the T_α's and the dimensions of the vector space components of D/t(D), we shall know D completely.

If R is a 1-dimensional Cohen-Macaulay ring with nonzero nilpotent elements, then we can still reduce the study of its torsion modules to the local case. We shall presently define Artinian R-modules and observe that they are torsion R-modules, and hence can be studied locally. For these reasons we shall restrict our attention in the rest of these notes to local, 1-dimensional Cohen-Macaulay rings. But before we do this we still need some results of a general nature.

Definitions. An ideal of a commutative ring is said to be *irreducible* if it is not the intersection of two properly larger ideals. An R-module is said to be *indecomposable* if it is not the direct sum of two properly smaller submodules.

The proof of the next theorem may be found in [9] and we shall not prove it here.

Theorem 4.5. *Let R be a commutative Noetherian ring. Then:*

(1) *An injective R-module C is indecomposable if and only if* $C = E(R/P)$, *where* $E(R/P)$ *is the injective envelope of* R/P, P *a prime ideal of* R.

(2) *An ideal* I *of* R *is irreducible if and only if* $I = \mathrm{Ann}_R(x)$, *where* x *is a nonzero element of an indecomposable injective* R-*module*.

(3) *Every injective* R-*module is a direct sum of indecomposable injective* R-*modules*.

(4) *Let* P *be a prime ideal of* R *and* $\hat{R}_P$ *the* P-*adic completion of* R_P. *Then* E(R/P) *is also the injective envelope over* R_P *and* $\hat{R}_P$ *of* R_P/PR_P. *Viewing* E(R/P) *as an* $\hat{R}_P$-*module gives rise to a natural isomorphism between* $\hat{R}_P$ *and* $\mathrm{Hom}_R(E(R/P),E(R/P))$.

(5) *If* P *is a prime ideal of* R, *then* $E = E(R/P) = \bigcup_n \mathrm{Ann}_E(P^n)$, *where* $\mathrm{Ann}_E(P^n) = \{x \in E \mid P^n x = 0\}$. *If* P *is a maximal ideal of* R, *then* $\mathrm{Ann}_E(P^n)$ *is a finitely generated* R-*module*.

(6) *If* I *is an ideal of* R, *and* $I = I_1 \cap \ldots \cap I_n$ *is an irredundant intersection of irreducible ideals* I_j, *then* $E(R/I) \cong E(R/I_1) \oplus \ldots \oplus E(R/I_n)$. *Furthermore*, I_j *is* P_j-*primary for some prime ideal* P_j *and* $E(R/I_j) \cong E(R/P_j)$.

(7) *Let* R *be a complete, Noetherian local ring with maximal ideal* M, *and* $E = E(R/M)$. *If* I *is an ideal of* R, *then* $\mathrm{Ann}_R(\mathrm{Ann}_E(I)) = I$; *and if* D *is a submodule of* E, *then* $\mathrm{Ann}_E(\mathrm{Ann}_R(D)) = D$.

Definitions: If R is a commutative ring, A an R-module, and $\mathfrak{a}$ a set of R-submodules of A, then $\mathfrak{a}$ is said to satisfy the *Ascending Chain Condition* (ACC) if for every ascending chain of modules in $\mathfrak{a}$: $A_1 \subset A_2 \subset \ldots \subset A_n \subset \ldots$ there is an index n_0 such that $A_n = A_{n_0}$ for all $n \geq n_0$. This is equivalent to the assertion that every subset of $\mathfrak{a}$ has a maximal element with respect to inclusion. If A has ACC on the set of all of its submodules, then A is said to be a *Noetherian* R-module. A Noetherian R-module is finitely generated.

In similar fashion we define the *Descending Chain Condition* (DCC) by reversing the inclusion relations. To say that $\mathfrak{a}$ satisfies DCC is equivalent to asserting that every subset of $\mathfrak{a}$ has a minimal element with respect to inclusion. If A has DCC on the set of all of its submodules, then A is said to be an *Artinian* R-module. An Artinian module over an integral domain or a 1-dimensional Noetherian Cohen-Macaulay ring is a torsion R-module.

If an R-module is both Artinian and Noetherian, we shall say that it has *finite length* (in the classical sense).

The proof of the following theorem is obtained by a trivial modification of the proofs of [9, Theorem 4.2] and [9, Corollary

4.3] and shall not be repeated here.

Theorem 4.6. (Duality). *Let R be a commutative, Noetherian local ring with maximal ideal M, let $\hat{R}$ be the completion of R in the M-adic topology, and let $E = E(R/M)$ be the injective envelope of R/M. Let A be an R-module. Then A is an Artinian R-module if and only if $\mathrm{Hom}_R(A, E)$ is a Noetherian $\hat{R}$-module. In this case we have:*

$$A \cong \mathrm{Hom}_{\hat{R}}(\mathrm{Hom}_R(A,E),E).$$

In particular A is an Artinian R-module if and only if A is a submodule of E^n, where E^n is a direct sum of $n > 0$ copies of E.

Conversely, let B be an $\hat{R}$-module. Then B is a Noetherian $\hat{R}$-module if and only if $\mathrm{Hom}_{\hat{R}}(B,E)$ is an Artinian R-module. In this case, we have:

$$B \cong \mathrm{Hom}_R(\mathrm{Hom}_{\hat{R}}(B,E),E).$$

The next theorem is a generalization of [13, Cor. 4.2], and the proof given here is due to Hamsher [4].

Theorem 4.7. *Let R be a commutative, quasi-local ring. Then K is an indecomposable R-module.*

Proof. Suppose that $K = A/R \oplus B/R$, where A and B are R-submodules of Q containing R and $A \neq Q$ and $B \neq Q$. Let c be a regular element and non-unit of R. Then $(c^{-1} + R) = x + y$, where $x \in A/R$ and $y \in B/R$. Therefore, $cx = 0 = cy$, and hence there exist elements r and t in R such that $x = rc^{-1} + R$ and $y = tc^{-1} + R$. But then $c^{-1} - (r + t)c^{-1} \in R$, and thus $1 - (r + t) \in Rc$. Since c is not a unit and R is a quasi-local ring, it follows that either r or t is a unit. Without loss of generality we can assume that r is a unit. But then $(c^{-1} + R) = r^{-1}x \in A/R$. Thus we have shown that if c is a regular element of R, then either $c^{-1} \in A$ or $c^{-1} \in B$.

Because $A \neq Q$ and $B \neq Q$, there exist elements c and d in R that are regular elements and not units of R such that $c^{-1} \in A$ and $d^{-1} \in B$. Now $c^{-1}d^{-1} \in A \cup B$, and hence we can assume that $c^{-1}d^{-1} \in A$. Thus $d^{-1} = c(c^{-1}d^{-1}) \in A \cap B = R$, contradicting the fact that d is not a unit of R. Therefore, K is an indecomposable R-module.

<u>Theorem</u> 4.8. <u>Let</u> R <u>be</u> <u>a</u> <u>Noetherian</u> <u>domain</u> <u>of</u> <u>Krull</u> <u>dimension</u> 1, M <u>a</u> <u>maximal</u> <u>ideal</u> <u>of</u> R, <u>and</u> $\{M_\alpha\}$ <u>a</u> <u>collection</u> <u>of</u> <u>maximal</u> <u>ideals</u> <u>of</u> R <u>different</u> <u>from</u> M. <u>Then</u> $\bigcap_\alpha R_{M_\alpha} \otimes_R R_M \cong Q$ <u>and</u> <u>hence</u> $(\bigcap_\alpha R_{M_\alpha}) + R_M = Q$.

<u>Proof</u>. If $x \in Q$, then $x = a/b$, where $a, b \in R$ and $b \neq 0$. Since b is contained in only a finite number of maximal ideals of R, we see that $x \in R_{M_\alpha}$ for all but a finite number of M_α's. Thus we can define an R-homomorphism $\varphi : Q \to \sum_\alpha \oplus K_{M_\alpha}$ by $\varphi(x) = \sum_\alpha (x + R_{M_\alpha})$ for all $x \in Q$. It is obvious that $\operatorname{Ker} \varphi = \bigcap_\alpha R_{M_\alpha}$, and thus we have a monomorphism $0 \to Q/(\bigcap_\alpha R_{M_\alpha}) \to \sum_\alpha \oplus K_{M_\alpha}$. By Theorem 4.1, K_{M_α} is isomorphic to the M_α-primary component of K and hence $K_{M_\alpha} \otimes_R R_M = 0$. Thus we have $(\sum_\alpha \oplus K_{M_\alpha}) \otimes_R R_M = 0$, from which it follows that $Q/(\bigcap_\alpha R_{M_\alpha}) \otimes_R R_M = 0$. Therefore, $(\bigcap_\alpha R_{M_\alpha}) \otimes_R R_M \cong Q$. It follows from this that $[(\bigcap_\alpha R_{M_\alpha}) + R_M]_N = Q$ for every maximal ideal N of R, and thus $(\bigcap_\alpha R_{M_\alpha}) + R_M = Q$.

CHAPTER V

ARTINIAN DIVISIBLE MODULES

Throughout the remainder of these notes R will be a Noetherian, local, 1-dimensional, Cohen-Macaulay ring with maximal ideal M.

Theorem 5.0. The following statements are true.

(1) *The* R-*topology and the* M-*adic topology on an* R-*module are the same.*

(2) H *is complete and is the completion of* R *in the* M-*adic topology.*

(3) H *is a complete, Noetherian, local,* 1-*dimensional Cohen-Macaulay ring with maximal ideal* HM.

(4) H *is a faithfully flat* R-*module.*

(5) *If* A *is a finitely generated* R-*module, then* $H \otimes_R A$ *is the completion of* A *in the* R (*or* M-*adic*) *topology.*

Proof. (1) A regular ideal of R is an M-primary ideal of R, and hence contains a power of M. It follows easily from this that the R-topology and the M-adic topology on an R-module are the same.

(2) If I is a regular ideal of R, then by Theorem 2.8, we have $H/HI \cong R/I$. Thus the completion of H in the R-topology is equal to the completion of R in the R-topology; that is, H is complete in the R-topology. Therefore H is complete in the M-adic topology, and is the M-adic completion of R.

(3) Statements (3), (4), and (5) could of course be established by a reference to general theorems in any standard textbook on Noetherian rings. However, in the 1-dimensional case we are considering, they follow readily from theorems we have already proved. Hence for the sake of completeness, we shall sketch the proofs here. By Corollary 2.5 and Theorem 2.8, H is a commutative, quasi-local ring with maximal ideal HM. Clearly the (HM)-adic topology on H is equal to the M-adic topology on H, and thus H is complete in the

(HM)-adic topology. If b is a regular element of M, then Rb contains a power of M and hence $\bigcap_n (HM)^n = \bigcap_n Hb^n$. But $\bigcap_n Hb^n$ is a divisible R-submodule of H, and thus $\bigcap_n (HM)^n = 0$. HM is a finitely generated ideal of H, and is generated by elements $b_1,\ldots,b_k$.

Let G(H) be the graded ring of H with respect to HM; that is, $G(H) = \sum_{n=0}^{\infty} \oplus (HM)^n/(HM)^{n+1}$ with multiplication defined in the obvious way. If we let $\bar{b}_i = b_i + (HM)^2$, then $G(H) = (H/HM)[\bar{b}_1,\ldots,\bar{b}_n]$, and hence G(H) is a commutative, Noetherian ring. If h is a nonzero element of H, then there exists an integer $n \geq 0$ such that $h \in (HM)^n$ but $h \notin (HM)^{n+1}$. If we let $\bar{h} = h + (HM)^{n+1}$ in G(H), then $\bar{h}$ is called the <u>leading form</u> of h.

If J is a nonzero ideal of H, let $\mathcal{J}$ be the ideal of G(H) generated by the leading forms of the nonzero elements of J. Since G(H) is a Noetherian ring, there is a finite set of elements $h_1,\ldots,h_t$ in J whose leading forms generate $\mathcal{J}$ over G(H). Since H is complete in the (HM)-adic topology and $\bigcap_n (HM)^n = 0$, it is not hard to show that $h_1,\ldots,h_t$ generate J over H. Thus H is a Noetherian ring.

Let b be a regular element of M. Since Hb contains a power of HM, Hb is an HM-primary ideal. It follows from the Krull principal ideal theorem [18, Ch. III, Theorem 6] that H has Krull dimension 1. Therefore, H is a complete, local, 1-dimensional, Cohen-Macaulay ring.

(4) and (5). Since $hd_R Q = 1$ by Theorem 4.2, these statements are true by Corollary 2.6 (3).

<u>Remarks</u>. We shall now recapitulate some of the results of the earlier chapters as they apply to a local, 1-dimensional, Cohen-Macaulay ring R. As before, Q is the full ring of quotients of R and K = Q/R. We have $hd_R Q = 1$ and $hd_R K = 1$. H is the completion of R is the R-topology, and $H \cong Hom_R(K,K)$. H/R is a torsion-free and divisible R-module isomorphic to $Ext_R^1(Q,R)$.

Every divisible R-module is h-divisible. If R has no nonzero nilpotent elements, then the torsion submodule of a divisible R-module is a direct summand. If B is an R-module, then h(B) is its divisible submodule, and B/h(B) is reduced. If B is a torsion R-module, then $h(B) \cong K \otimes_R \mathrm{Hom}_R(K,B)$. If A is a torsion-free R-module, then its completion is a cotorsion R-module and is isomorphic to $\mathrm{Hom}_R(K, K \otimes_R A)$.

We shall denote the maximal ideal of R by M. An ideal of R is regular if and only if it is an M-primary ideal. The R-topology and the M-adic topology on an R-module are the same. H is a complete, Noetherian, local, 1-dimensional, Cohen-Macaulay ring with maximal ideal HM and $H/HM \cong R/M$. H is a faithfully flat R-module. If A is a finitely generated R-module, then its completion is isomorphic to $H \otimes_R A$.

If E = E(R/M) is the injective envelope of R/M over R, then E is also the injective envelope of H/HM over H. By Theorem 4.5 we have $\mathrm{Hom}_R(E,E) \cong H$. We also have available the duality of Theorem 4.6 between Noetherian H-modules and Artinian R-modules given by the functor $\mathrm{Hom}_R(\cdot,E)$. An Artinian R-module is an Artinian H-module, and the converse is also true. Artinian R-modules are torsion R-modules.

Definitions. If A is an R-module and I is an ideal of R we shall define the annihilator of A in R by $\mathrm{Ann}_R A = \{r \in R \mid rA = 0\}$ and the annihilator of I in A by $\mathrm{Ann}_A I = \{x \in A \mid Ix = 0\}$.

Theorem 5.1. *Let A be an Artinian R-module. Then A is reduced if and only if A is finitely generated (i.e., has finite length).*

Proof. By the Nakayama Lemma a finitely generated R-module is reduced. Conversely, suppose that A is reduced but not finitely generated. Since A is Artinian, it contains a submodule B that is minimal with respect to the property of not being finitely generated. Because B is reduced, there exists a regular element $r \in R$ such that

$B \neq rB$. By the minimality of B we see that rB is finitely generated. It follows that B/rB is not finitely generated.

Since B/rB is Artinian, we have $B/rB \subset E^n$ for some $n > 0$ by Theorem 4.6. Because Rr is an M-primary ideal of R, there exists an integer $k > 0$ such that $M^k \subset Rr$. Thus we have $B/rB \subset \mathrm{Ann}_{E^n} Rr \subset \mathrm{Ann}_{E^n} M^k$. But $\mathrm{Ann}_{E^n} M^k$ is a direct sum of n copies of $\mathrm{Ann}_E M^k$, and $\mathrm{Ann}_E M^k$ is a finitely generated R-module by Theorem 4.5. Therefore, B/rB is finitely generated, and this contradiction proves that A is finitely generated.

Corollary 5.2. *If A is an Artinian module, then A/h(A) is finitely generated and hence has finite length in the classical sense. Thus $A = h(A) + B$, where B is a finitely generated submodule of A.*

Proof. A/h(A) is Artinian and reduced, and hence is finitely generated by Theorem 5.1.

Theorem 5.3. *Let D be a divisible R-module, B a submodule of D, and C a finitely generated submodule of D. If $D = B + C$, then $D = B$.*

Proof. We have $D/B \cong C/(B \cap C)$, and hence D/B is both divisible and finitely generated. By the Nakayama Lemma, we see that $D = B$.

Theorem 5.4. *Given an exact sequence of R-modules:*

$$0 \to A \to B \to C \to 0$$

then

(1) *If A and C are reduced, B is also reduced.*

(2) *If B is reduced and A is Artinian, then C is reduced.*

Proof. If A and C are reduced, then clearly B is reduced. Hence suppose that B is reduced and that A is Artinian. Then

$\mathrm{Hom}_R(Q,B) = 0$ and hence we have an exact sequence:

$$0 \to \mathrm{Hom}_R(Q,C) \to \mathrm{Ext}^1_R(Q,A).$$

By Theorem 5.1, A has finite length, and thus $\mathrm{Ext}^1_R(Q,A)$ is a torsion module. On the other hand $\mathrm{Ext}^1_R(Q,A)$ is torsion-free and divisible. Thus $\mathrm{Ext}^1_R(Q,A) = 0$, and by the preceding exact sequence this implies that $\mathrm{Hom}_R(Q,C) = 0$. Hence C has no nonzero h-divisible submodules, and thus C is reduced.

Definitions. We will say that an R-module is a simple divisible module if it is a nonzero torsion, divisible module that has no proper nonzero divisible submodules.

We will say that a torsion divisible R-module D has a composition series of divisible modules if it has a chain of divisible submodules:

$$0 = D_0 \subset D_1 \subset D_2 \subset \dots \subset D_n = D,$$

such that D_i/D_{i-1} is a simple divisible R-module for $i = 1,\dots,n$.

The next theorem shows that the property of having a composition series of divisible modules is confined to the class of Artinian divisible modules. More surprisingly it also shows that either the ACC or the DCC on divisible submodules is separately equivalent to the apparently stronger condition of being Artinian.

Definition. We shall let K^n denote a direct sum of n copies of K. A similar statement holds for E^n.

Theorem 5.5. *If D is a torsion divisible module, then the following statements are equivalent:*

(1) *D has both DCC and ACC on divisible submodules.*

(2) *D has a composition series of divisible modules.*

(3) *D is a homomorphic image of K^n for some $n > 0$.*

(4) D _is a submodule of_ E^n _for some_ $n > 0$.

(5) D _is an Artinian module_.

(6) D _has_ ACC _on divisible submodules_.

(7) D _has_ DCC _on divisible submodules_.

Proof. (4) <==> (5). This equivalence has already been stated in Theorem 4.6.

(1) ==> (2). Since D has DCC on divisible submodules, D has a simple divisible submodule D_1. Clearly D/D_1 has DCC on divisible submodules and thus has a simple divisible submodule D_2/D_1. Now D_2 is a divisible submodule of D and we may proceed with D/D_2. Since D has ACC on divisible submodules, the process stops in a finite number of steps with a composition series of divisible submodules of D.

(2) ==> (3). Let $0 = D_0 \subset D_1 \subset \dots \subset D_n = D$ be a composition series of divisible submodules of D. By Corollary 1.3, $\mathrm{Hom}_R(K,D_1) \neq 0$; and since D_1 is a simple divisible module, every nonzero element of $\mathrm{Hom}_R(K,D_1)$ is an epimorphism. Suppose that for a given $i > 0$ we have found an epimorphism $g_i : K^i \to D_i$. Choose $x \in D_{i+1} - D_i$, and by Corollary 1.3 an R-homomorphism $f: K \to D_{i+1}$ such that $x \in f(K)$. Then $D_{i+1} = D_i + f(K)$, and using g_i and f we can obtain an epimorphism $g_{i+1} : K^{i+1} \to D_{i+1}$. Hence by induction there is an epimorphism g_n on K^n onto $D_n = D$.

(3) ==> (5). If we define $M^{-1} = \{q \in Q \mid qM \subset R\}$, then there is a regular element $b \in R$ such that $bM^{-1} \subset R$. Hence M^{-1}/R is a finitely generated submodule of K. It is easy to see that K is an essential extension of M^{-1}/R and thus $K \subset E(M^{-1}/R)$, the injective envelope of M^{-1}/R. Since M^{-1}/R is finitely generated, $E(M^{-1}/R)$ is a finite direct sum of copies of E by Theorem 4.5. Therefore, K (and hence also K^n) is contained in a finite direct sum of copies of E. Thus K^n is Artinian by Theorem 4.6, and hence every homomorphic image of K^n is also Artinian.

(5) ==> (6). By Theorem 4.6 we have $D \subset E^n$ for some $n > 0$. Let

$D_1 \subset D_2 \subset \dots \subset D_m \subset \dots$ be a chain of divisible submodules of D. Then $\mathrm{Hom}_R(K,D_1) \subset \mathrm{Hom}_R(K,D_2) \subset \dots \subset \mathrm{Hom}_R(K,D_m) \subset \dots$ is a chain of H-submodules of $\mathrm{Hom}_R(K,E^n)$ by Theorem 2.7. But K is an Artinian R-module by the proof of (3) ==> (5), and hence $\mathrm{Hom}_R(K,E^n)$ is a Noetherian H-module by Theorem 4.6. Thus there is an index m_0 such that $\mathrm{Hom}_R(K,D_m) = \mathrm{Hom}_R(K,D_{m_0})$ for all $m \geq m_0$. Therefore $D_m = D_{m_0}$ for all $m \geq m_0$ by Corollary 1.3, proving that D has ACC on divisible submodules.

(6) ==> (1). Since D has ACC on divisible submodules, there is a maximal Artinian divisible submodule A of D. If $x \in D$, then by Corollary 1.3 there is an R-homomorphism $f: K \to D$ such that $x \in f(K)$. Since K is Artinian by (3) ==> (5), we see that $A + f(K)$ is Artinian, and hence $f(K) \subset A$. Therefore, $A = D$.

(5) ==> (7). This is a trivial implication.

(7) ==> (5). It is a consequence of (3) ==> (5) and Corollary 1.3 that every torsion, divisible R-module is a sum of Artinian divisible R-modules. We assert that since D has DCC on divisible submodules, there is an Artinian divisible submodule B of D such that if A is any submodule of D satisfying $A \cap B = 0$, then A is reduced. Suppose that this assertion is false and choose a nonzero Artinian divisible submodule B_1 of D. Then there is a submodule A_1 of D such that $A_1 \cap B_1 = 0$ and A_1 is not reduced. Choose a nonzero Artinian divisible submodule B_2 of A_1. Then $B_1 \oplus B_2$ is a nonzero Artinian divisible submodule of D. Continuing in this way we see that we can construct a divisible submodule of D that is an infinite direct sum. This contradicts the fact that D has DCC on divisible submodules, and hence we have established the existence of a submodule B of D with the desired properties.

We denote the injective envelopes of B and D by E(B) and E(D), respectively; and we can assume that E(B) is a direct summand of E(D). Thus there is a submodule C of E(D) such that $E(D) = E(B) \oplus C$.

If we let $A = D \cap C$, then $A \cap B = 0$, and hence A is reduced. Since D/A is isomorphic to a submodule of E(B), and E(B) is Artinian by Theorem 4.6, we see that D/A is an Artinian divisible module.

Since D is a sum of Artinian divisible R-modules as we have remarked, it is sufficient to prove that D has ACC on Artinian divisible submodules in order to prove that D is Artinian. Let $D_1 \subset D_2 \subset \ldots \subset D_n \subset \ldots$ be an ascending chain of Artinian divisible submodules of D. Since $(D_i + A)/A \cong D_i/(D_i \cap A)$ is divisible we see that $\{(D_i + A)/A\}$ is an ascending chain of Artinian divisible submodules of D/A. However, D/A is Artinian and hence has ACC on divisible submodules by (5) ⟹ (6). Hence there is an integer n_0 such that $D_n + A = D_{n_0} + A$ for all $n \geq n_0$. From this it follows immediately that $D_n = D_{n_0} + (D_n \cap A)$ for all $n \geq n_0$. Therefore, $D_n/D_{n_0} \cong (D_n \cap A)/(D_{n_0} \cap A)$. Since $D_n \cap A$ is Artinian and reduced, $(D_n \cap A)/(D_{n_0} \cap A)$ is also reduced by Theorem 5.4. Hence D_n/D_{n_0} is both divisible and reduced, and we have $D_n = D_{n_0}$ for all $n \geq n_0$. Thus D is an Artinian module.

Corollary 5.6. *If a torsion divisible R-module D has a reduced submodule A such that D/A is Artinian, then D is Artinian.*

Proof. This assertion was proved as part of proving that (7) ⟹ (5) in the preceding theorem.

Corollary 5.7. *Let D be a torsion divisible R-module. Then D is a simple divisible R-module if and only if every proper R-submodule of D is finitely generated. In this case D is an Artinian R-module.*

Proof. If D is a simple divisible R-module, then D satisfies DCC on divisible submodules in a trivial way, and hence D is Artinian by Theorem 5.5. Since a proper submodule of D is reduced, it is finitely generated by Theorem 5.1.

Definition. We shall say that two R-modules are equivalent if each is a homomorphic image of the other. This is clearly an equivalence relation on the class of R-modules. If A and B are equivalent R-modules, we shall denote this by $A \sim B$, and we shall denote the class of R-modules equivalent to A by [A].

Lemma 5.8. If A is an Artinian divisible R-module and B is a reduced submodule of A, then A/B is equivalent to A. Thus if A is a simple divisible R-module, every nonzero homomorphic image of A is a simple divisible module equivalent to A.

Proof. Let $D = A/B$; then we have an exact sequence:

$$(*) \qquad 0 \to \mathrm{Hom}_R(D,B) \to \mathrm{Hom}_R(D,A) \to \mathrm{Hom}_R(D,D) \to \mathrm{Ext}^1_R(D,B)$$

By Theorem 5.1, B has finite length, and thus $\mathrm{Ext}^1_R(D,B)$ has a regular annihilator. Let $r \in R$ be a regular element of the annihilator of $\mathrm{Ext}^1_R(D,B)$, let I be the identity mapping on D, and let $\pi: A \to D$ be the canonical homomorphism. Then by exact sequence (*) we see that there exists $g \in \mathrm{Hom}_R(D,A)$ such that $\pi g = rI$. Thus πg is an epimorphism. It follows that $A = B + g(D)$. Since B is finitely generated we have $A = g(D)$ by Theorem 5.3, and hence g is an epimorphism. Therefore, we have proved that A is equivalent to D.

If A is a simple divisible R-module, then A is Artinian by Corollary 5.7, and the second statement of the Lemma follows from the first and Corollary 5.7.

Lemma 5.9. Let A be an R-module and $B \subset C$ submodules of A such that C/B is a simple divisible R-module. If S is a submodule of A such that $S \cap C$ is Artinian and reduced, then $(S + C)/(S + B)$ is a simple divisible R-module equivalent to C/B.

Proof. Since $B \subset (S + B) \cap C$ and $(S + C)/(S + B) \cong C/((S + B) \cap C)$, we see that $(S + C)/(S + B)$ is a homomorphic image of C/B and is either 0 or a simple divisible module equivalent to C/B according to Lemma 5.8. Therefore it is sufficient to assume that $S + C = S + B$ and arrive at a contradiction. But in this case we have $C = B + (S \cap C)$, and hence $C/B \cong (S \cap C)/(S \cap B)$. Since $S \cap C$ is finitely generated by Theorem 5.1, and C/B is a nonzero divisible module, we have our desired contradiction.

Definition. If A is an Artinian divisible R-module, we shall say that two composition series of divisible submodules of A are equivalent if there is a one-to-one correspondence between the sets of factor modules of the two series such that corresponding factor modules are equivalent. Of course equivalent composition series then have the same number of terms in their series.

Given two descending chains of submodules of A, we shall say that one of the chains is a refinement of the other if it contains every term of the other chain somewhere in its sequence.

Theorem 5.10 (A Jordan-Hölder type of Theorem). *Let A be an Artinian divisible R-module. Then*

(1) *A has a composition series of divisible R-modules.*

(2) *Any two composition series for A are equivalent.*

(3) *Any chain of divisible submodules of A can be refined to a composition series.*

Proof. (1) This is part of Theorem 5.5.

(2) Let $0 = A_0 \subset A_1 \subset \cdots \subset A_n = A$ be a composition series of divisible submodules of A, and let B be a simple divisible submodule of A. There is an integer i such that $B \subset A_i$, but $B \not\subset A_{i-1}$. Because of an obvious induction argument it is sufficient to prove that

$$0 \subset B \subset B + A_1 \subset \cdots \subset B + A_{i-1} \subset A_{i+1} \subset \cdots \subset A_n = A$$

is a composition series of divisible submodules of A equivalent to the given series in order to complete the proof of (2).

Now $A_{i-1} \subsetneq B + A_{i-1} \subset A_i$, and $B + A_{i-1}$ is a divisible submodule of A_i. Since A_i/A_{i-1} is a simple divisible R-module, there are no divisible modules properly between A_{i-1} and A_i. Therefore, $B + A_{i-1} = A_i$ and hence $A_i/A_{i-1} = (B + A_{i-1})/A_{i-1} \cong B/(B \cap A_{i-1})$ is equivalent to B by Lemma 5.8. On the other hand for $j \leq i - 1$ we see that $(B + A_j)/(B + A_{j-1})$ is equivalent to A_j/A_{j-1} by Lemma 5.9 and we have proved our assertion.

(3) This is an immediate consequence of Theorem 5.5.

Definition. If A is an Artinian R-module, then according to Theorem 5.10 every composition series of divisible submodules of h(A) has the same number of terms. Thus we can unambiguously define L(A) to be the number of terms in a composition series of divisible submodules of h(A). It can be immediately seen that L(A) = 0 if and only if A is reduced.

Theorem 5.11. *Let* $0 \to A \to B \to C \to 0$ *be an exact sequence of Artinian R-modules. Then:*

$$L(B) = L(A) + L(C).$$

Proof. We shall assume that C = B/A and that $\pi : B \to C$ is the canonical map. By Corollary 5.2, $B = h(B) + B_1$, where B_1 is a finitely generated R-module. We then have $C = \pi(h(B)) + \pi(B_1)$, and it follows easily from Theorem 5.3 that $\pi h(B) = h(C)$. Thus we have an exact sequence

$$0 \to h(B) \cap A \to h(B) \to h(C) \to 0.$$

Now L(B) = L(h(B)) and L(C) = L(h(C)); and since $h(A) \subset h(B) \cap A$ we have $L(h(B) \cap A) = L(A)$. Thus without loss of generality we can assume that B is a divisible R-module.

Now we have an exact sequence:

$$0 \to A/h(A) \to B/h(A) \to C \to 0.$$

Since $A/h(A)$ is reduced, we have $L(A/h(A)) = 0$. It is an immediate consequence of Theorem 5.10 that because $h(A)$ is divisible we have $L(B/h(A)) = L(B) - L(h(A)) = L(B) - L(A)$. Thus without loss of generality, we can assume that A is reduced. It then follows from Lemma 5.9 that the image under π of a composition series of divisible submodules of B is a composition series of divisible submodules of C. We then have $L(B) = L(C) = L(C) + L(A)$, since $L(A) = 0$.

Corollary 5.12. *If A and B are equivalent Artinian R-modules, then* $L(A) = L(B)$.

Proof. Since B is a homomorphic image of A, we have $L(B) \leq L(A)$; and since A is a homomorphic image of B we have $L(A) \leq L(B)$.

Corollary 5.13. *If A and B are Artinian divisible R-modules, then A and B are equivalent if and only if there exists a homomorphism f of A onto B such that Ker f is reduced.*

Proof. If such a homomorphism exists, then A and B are equivalent by Lemma 5.8. On the other hand, if A and B are equivalent, we have an exact sequence:

$$0 \to \operatorname{Ker} f \to A \xrightarrow{f} B \to 0.$$

By Theorem 5.11 we have $L(A) = L(B) + L(\operatorname{Ker} f)$, and by Corollary 5.12 we have $L(A) = L(B)$. Thus $L(\operatorname{Ker} f) = 0$, and hence Ker f is reduced.

Corollary 5.14. *If A and B are equivalent Artinian divisible R-modules, then A and B have equivalent composition series.*

Proof. Corollary 5.13 and Lemma 5.9.

Remarks. The converse of Corollary 5.14 is not true in general. We shall prove in a later theorem that the converse of Corollary 5.14

is true if and only if the integral closure of R in Q is a finitely generated R-module.

Theorem 5.15. Let $f \in \mathrm{Hom}_R(K,K)$ *and let* Ker $f = C/R$. *Then the following statements are equivalent:*

(1) *f is an epimorphism*

(2) *C is a finitely generated R-module.*

(c) $C \cong R$.

Thus if B is an R-submodule of Q, then $Q/B \cong K$ *if and only if* $B \cong R$.

Proof. (1) $\Longleftrightarrow$ (2). Since $L(\mathrm{Ker}\ f) = L(K) - L(f(K))$, we see that Ker f is reduced if and only if f is an epimorphism. But K is Artinian by Theorem 5.5 and hence by Theorem 5.1, Ker f is reduced if and only if Ker f is finitely generated.

(2) $\Longrightarrow$ (3). Since Ker $f = C/R$, we have $f(K) \cong Q/C$. But $f(K) = K$ by (2) $\Longrightarrow$ (1); and since Q/C is a torsion R-module we have $Q/C \cong K \otimes_R C$ by Theorem 1.1. Thus $K \cong K \otimes_R C$, and hence $H \cong \mathrm{Hom}_R(K,K) \cong \mathrm{Hom}_R(K, K \otimes_R C)$. Therefore the completion of C is isomorphic to H by Theorem 2.2. Since C is isomorphic to a regular ideal of R, the completion of C is isomorphic to $H \otimes_R C$ by Theorem 2.8. Thus $H \cong H \otimes_R C$, and by Corollary 2.6, C is a projective R-module. However, R is a local ring, and thus projective R-modules are free. Since C is an R-submodule of Q, it is an indecomposable R-module. Thus we have $C \cong R$.

(3) $\Longrightarrow$ (2). This implication is trivial.

Suppose that B is an R-submodule of Q such that $Q/B \cong K$. Without loss of generality we can assume that $R \subset B$. Thus we have an epimorphism of K onto Q/B with kernel B/R, and hence an epimorphism of K onto K with kernel B/R. By (1) $\Longrightarrow$ (3) we see that $B \cong R$. Conversely, if $B \cong R$, then it can be immediately seen that $Q/B \cong K$.

Theorem 5.16. If A is a nonzero Artinian R-module, then A does not contain a proper submodule isomorphic to itself. Thus every

<u>monomorphism of A into itself is an isomorphism.</u>

<u>Proof.</u> Suppose that B is a proper submodule of A and that f is an isomorphism of A onto B. Now $h(B) \subset h(A)$, and $L(h(B)) = L(B) = L(A) = L(h(A))$. Thus we have $h(B) = h(A)$, and $f(h(A)) = h(B) = h(A)$. Hence f induces an isomorphism of $A/h(A)$ onto a proper submodule of itself, namely $B/h(A)$. Therefore, by Corollary 5.2 we can assume without loss of generality that A has finite length in the classical sense.

We shall let $\mathcal{L}(A)$ denote the length of a composition series of A whose factors are ordinary simple R-modules isomorphic to R/M. We then have $\mathcal{L}(A) = \mathcal{L}(B) + \mathcal{L}(A/B)$. But $\mathcal{L}(A) = \mathcal{L}(B)$, since A and B are isomorphic. Hence $\mathcal{L}(A/B) = 0$, showing that B = A. This contradiction proves that A cannot be isomorphic to B.

<u>Remarks.</u> The dual assertion to Theorem 5.16 is false in general. That is, there do exist isomorphisms of Artinian modules onto proper factor modules of themselves. As an example, we may let A = K. Let c be regular element of R that is not a unit in R and define $B = Rc^{-1}/R$. Then A/B is a proper factor module of A that is isomorphic to A.

<u>Theorem 5.17. Let A be a torsion divisible R-module. Then A is a sum of Artinian divisible R-modules.</u>

<u>Proof.</u> This is an immediate consequence of Theorem 5.5 and Corollary 1.3.

CHAPTER VI

STRONGLY UNRAMIFIED RING EXTENSIONS

Throughout this chapter R will be a Noetherian, local, 1-dimensional, Cohen-Macaulay ring with maximal ideal M.

Definition. If A is a commutative ring extension of R, we shall say that A is strongly unramified over R if AM is a regular ideal of A, AM contains every regular ideal of A, and $A/AM \cong R/M$. Of course, AM is then a maximal ideal of A, $AM \cap R = M$, and $AM + R = A$. We note that A need not be a quasi-local ring. It is clear, however, that if A is strongly unramified over R, then $A_{(AM)}$ is an unramified extension of R in the usual sense.

Theorem 6.0. *H is a compatible ring extension of R and is strongly unramified over R. There is a one-to-one correspondence between the set of regular (i.e. M-primary) ideals I of R and the set of regular (i.e. HM-primary) ideals J of H given by $HI = J$ and $J \cap R = I$. Thus $HI \cap R = I$ and $J = H(J \cap R)$; moreover, $H/HI \cong R/I$ and $(HI)/I$ is torsion-free and divisible.*

Proof. Since R is a reduced R-module, we see from Corollary 2.5 that H is a torsion-free R-module and that H/R is a divisible R-module. Let J be a regular ideal of H. By Theorem 5.0, H is a local, 1-dimensional Cohen-Macaulay ring with maximal ideal HM. It follows that J is an HM-primary ideal of H and hence there exists an integer $n > 0$ such that $(HM)^n \subset J$. Hence $M^n \subset (HM)^n \cap R \subset J \cap R$; and thus $J \cap R$ is a regular ideal of R. Therefore, H is a compatible ring extension of R. By Theorem 2.8, $H/HM \cong R/M$, and so H is strongly unramified over R. The one-to-one correspondence of the theorem, and the remaining statements now follow directly from what has already been proved and Theorem 2.8.

The next theorem shows that there are many other compatible, strongly unramified extensions of R; and these play a big role in the theory of Artinian divisible R-modules.

Theorem 6.1. Let A *be an* R*-module such that* $R \subset A \subsetneq Q$. *(We do not assume in advance that* A *is a ring extension of* R*). Then the following statements are equivalent:*

(1) A/R *is a divisible* R*-module.*

(2) A *is a compatible ring extension of* R.

(3) A *is a strongly unramified ring extension of* R.

If A *is a strongly unramified ring extension of* R, *then all of the statements of Theorem 3.6 are true for* R. *Of particular importance are the following:*

(a) *If* S *is an* R*-module such that* $A \subset S \subset Q$, *then* S *is an* A*-module.*

(b) *A finitely generated torsion* A*-module has finite length as an* R*-module, and the same finite length over* A.

(c) *If* J *is a regular ideal of* A, *then* $J \subset AM$, $A = J + R$, $A/J \cong R/(J \cap R)$, *and* $J = A(J \cap R)$.

(d) *If* R *is an integral domain, then* A *is a Noetherian local domain of Krull dimension 1.*

Proof. The equivalence of (1) and (2) and the implication (2) $\Longrightarrow$ (3) as well as the statements at the end of the theorem are given by Corollary 3.5 and Theorem 3.6.

(3) $\Longrightarrow$ (1). If b is a regular element of M, then there is an integer $n > 0$ such that $M^n \subset Rb$. Since $A = AM + R$, we have $M(A/R) = A/R$ from which it follows that $b(A/R) = A/R$. Hence A/R is a divisible R-module.

Remarks. We can not overemphasize the importance of the statement in Theorem 6.1 that if R is a Noetherian local domain of Krull dimension 1, and A is a strongly unramified ring extension of R in Q, then A is also a Noetherian local domain of Krull dimension 1.

This fact greatly simplifies the proofs of theorems about local 1-dimensional domains and explains the difference between Corollary 6.2 and Theorem 6.9.

With the aid of Theorem 6.1 we shall give a short direct proof of the Theorem of Krull-Akizuki. (For a different proof, see [17, Theorem 33.2].

Corollary 6.2. (Theorem of Krull-Akizuki). Let R be a Noetherian domain of Krull dimension 1 with quotient field Q, let L be a finite algebraic field extension of Q, and let T be a ring (not a field) such that $R \subset T \subset L$. Then T is a Noetherian domain of Krull dimension 1; and if J is a nonzero ideal of T, then T/J is a module of finite length over R. If R is a local domain then T is a semi-local domain.

Proof. By taking a finite integral extension of R we may assume that $L = Q$. To prove the theorem it is sufficient to prove that if J is a nonzero ideal of T, then T/J is a module of finite length over R. Let $I = J \cap R$; then I is a nonzero ideal of R, and hence I is contained in only a finite number of maximal ideals of R. Thus $J_M = T_M$ for all but a finite number of maximal ideals M of R. By Theorem 4.1, $T/J \cong \Sigma \oplus T_M/J_M$, where M ranges over all maximal ideals of R. As we have just seen, this sum has only finitely many terms and thus it is sufficient to prove that T_M/J_M has finite length over R_M. Thus we may assume that R is a local domain.

Let $A/R = h(T/R)$; then by Theorem 6.1, A is a strongly unramified extension of R and T/J has finite length over R if and only if it has finite length over A. Hence we may assume that $R = A$. Thus T/R is a reduced R-submodule of K. Since K is an Artinian R-module by Theorem 5.5, we see that T/R has finite length by Theorem 5.1. It follows that $T/(J \cap R)$ has finite length. Because T/J is an R-homomorphic image of $T/(J \cap R)$, we see that T/J is a module of finite length over R.

Theorem 6.3. (Independence of strongly unramified extensions). *Assume that* R *is a Noetherian local domain of Krull dimension* 1. *Let* $A_1,\ldots,A_n$ *be strongly unramified extensions of* R *in* Q, N_i *the maximal ideal of* A_i, $S = A_1 \cap \ldots \cap A_n$ *and* $M_i = N_i \cap S$. *If* $A_i + A_j = Q$ *for* $i \neq j$, *then*:

(1) $M_1,\ldots,M_n$ *are the only maximal ideals of* S; *they are all distinct and* $S_{M_i} = A_i$.

(2) $Q/S \cong Q/A_1 \oplus \ldots \oplus Q/A_N$ *and* $A_i + \bigcap_{j\neq i} A_j = Q$.

Proof. Let $T = A_2 \cap \ldots \cap A_n$, and $P_j = N_j \cap T$ for $j \geq 2$. Then by induction on n, the P_j's are the only maximal ideals of T, and we have $T_{P_j} = A_j$ and $Q/T \cong Q/A_2 \oplus \ldots \oplus Q/A_n$. By Theorem 3.6 we have $A_1 + T = A_1T$, and thus $A_1 + T$ is a ring. In fact $A_1 + T = Q$. For suppose that this is not the case. Then by localizing $A_1 + T$ at a maximal ideal we obtain a 1-dimensional local ring U that contains both A_1 and T. Let N be the maximal ideal of U and $P = N \cap T$. Then P is a maximal ideal of T by Corollary 6.2, and hence $P = P_j$ for some $j \geq 2$. But then $A_j = T_{P_j} \subset U$, and thus $Q = A_1 + A_j \subset U$. This contradiction shows that $A_1 + T = Q$. Now $S = A_1 \cap T$, and thus
$Q/S = (T + A_1)/S = T/S \oplus A_1/S \cong (T + A_1)/A_1 \oplus (T + A_1)/T = Q/A_1$
$\oplus\, Q/T \cong Q/A_1 \oplus Q/A_2 \oplus \ldots \oplus Q/A_n$. Thus we have proved statement (2).

By Corollary 6.2, S is a Noetherian semi-local domain of Krull dimension 1, and thus each M_i is a maximal ideal of S. We have $Q/S_{M_1} \cong (Q/S)_{M_1} \cong (Q/A_1)_{M_1} \oplus \ldots \oplus (Q/A_n)_{M_1}$. However, Q/S_{M_1} is an indecomposable S_{M_1}-module by Theorem 4.7 and $(Q/A_1)_{M_1} = (Q/A_1)$. Therefore, we have $(A_j)_{M_1} = Q$ for all $j \geq 2$. This shows that $M_1 \not\subset N_j$ for $j \geq 2$, and hence $M_1 \neq M_j$ for $j \geq 2$. Thus we have proved that the M_i's are all distinct.

Since $S_{M_i} \subset A_i$ for all $i = 1,\ldots,n$, we have

$S \subset S_{M_1} \cap \ldots \cap S_{M_n} \subset A_1 \cap \ldots \cap A_n = S$. Therefore $S = S_{M_1} \cap \ldots \cap S_{M_n}$. If P is a maximal ideal of S different from the M_i's then $(S_{M_1} \cap \ldots \cap S_{M_n})_P = Q$ by Theorem 4.8. But then $S_P = Q$, and this contradiction shows that the M_i's are the only maximal ideals of S.

We have $S_{M_1} + (S_{M_2} \cap \ldots \cap S_{M_n}) = Q$ by Theorem 4.8. Thus if $x \in A_1$, we have $x = y + z$, where $y \in S_{M_1}$ and $z \in (S_{M_2} \cap \ldots \cap S_{M_n})$. Thus $z = x - y \in (A_1 + S_{M_1}) = A_1$, and hence $z \in A_1 \cap S_{M_2} \cap \ldots \cap S_{M_n} \subset A_1 \cap \ldots \cap A_n = S$. Therefore, $x \in S_{M_1} + S = S_{M_1}$, and we have shown that $A_1 = S_{M_1}$. By similar arguments we have $A_i = S_{M_i}$ for all $i = 1,\ldots,n$.

<u>Remarks</u>. The necessity of the condition $Q = A_i + A_j$ for $i \neq j$ in Theorem 6.3 is demonstrated by the fact that if $A_1 + \ldots + A_n \neq Q$, then $S = A_1 \cap \ldots \cap A_n$ is a local ring. The similarity of Theorem 6.3 with an analogous theorem concerning independent valuation rings [17, Theorem 11.18] will of course not go unnoticed by the careful reader.

<u>Theorem 6.4</u>. <u>Let</u> S <u>be</u> <u>a</u> <u>ring extension of</u> R <u>in</u> Q $(S \neq Q)$, <u>let</u> $B/R = h(S/R)$, <u>and let</u> $H(S)$ <u>be the completion of</u> S <u>in the</u> S-<u>topology</u>. <u>Then we have an exact sequence</u>:

$$0 \to \mathrm{Hom}_R(K,B/R) \to H \xrightarrow{\varphi} H(S) \to S/B \to 0$$

<u>where</u> φ <u>is a ring homomorphism</u>. <u>Furthermore</u>, S/B <u>has finite length as an</u> R-<u>module</u>. <u>Therefore</u> H(S) <u>is a finitely generated</u> H-<u>module</u>, <u>and hence is a complete</u>, <u>semi-local</u>, <u>Noetherian</u>, 1-<u>dimensional Cohen-Macaulay ring</u>.

<u>Proof</u>. It is clear that $\mathrm{Hom}_R(K,B/R) \cong \mathrm{Hom}_R(K,S/R)$. We have an exact sequence:

$$\mathrm{Ext}^1_R(K,B/R) \to \mathrm{Ext}^1_R(K,S/R) \to \mathrm{Ext}^1_R(K,S/B) \to \mathrm{Ext}^2_R(K,B/R).$$

Because $hd_R K = 1$, the end terms of this sequence are 0, and thus $Ext_R^1(K,S/R) \cong Ext_R^1(K,S/B)$. Now S/B is a reduced, Artinian R-module and hence has finite length by Theorem 5.1. Therefore, $Ext_R^1(K,S/B) \cong S/B$ by Theorem 1.1. The theorem now follows immediately from Theorem 2.9.

Lemma 6.5. *If I is an ideal of H, then H/I is a torsion-free R-module if and only if I is an unmixed ideal of rank 0 in H.*

Proof. Suppose that $I = J_1 \cap \dots \cap J_t$ is a normal decomposition of I in H, where J_i is P_i-primary and P_i is a prime ideal of rank 0 in H. If H/I is not a torsion-free R-module, there exists $h \in H - I$ and r regular in R such that $rh \in I$. But there exists an index i such that $h \notin J_i$, and hence $rh \in J_i$ implies that $r \in P_i$. However, $P_i \cap R$ has no regular elements, and this contradiction proves that H/I is a torsion-free R-module.

Conversely, suppose that H/I is a torsion-free R-module. If I is not unmixed of rank 0 in H, then HM belongs to I; and hence there exists $h \in H - I$ such that $HMh \subset I$. But then $Mh \subset I$, and H/I is not torsion-free over R. This contradiction proves that I is unmixed of rank 0 in H.

Definition. If I is an unmixed ideal of rank 0 in H, then H/I is a torsion-free R-module by Lemma 6.5, and hence by Theorem 1.1 we have a canonical monomorphism: $K \otimes_R I \to K \otimes_R H$. By Theorem 2.7 we can identify $K \otimes_R H$ with K, and from now on we shall consistently identify $K \otimes_R I$ with its image in K. Thus if $x \in K$ and $f \in I$, we shall identify $x \otimes f$ with $f(x)$.

Theorem 6.6. *There is a one-to-one, order preserving, correspondence between the set of unmixed ideals* $\{I\}$ *of rank 0 in H and the set of proper divisible submodules* $\{B/R\}$ *of K given by* $B/R = K \otimes_R I$ *and* $I = Hom_R(K,B/R)$.

If I *and* B/R *correspond, then:*

(1) B *is a strongly unramified ring extension of* R.

(2) H/I *is isomorphic as a ring to* H(B), *the completion of* B.

(3) $\mathrm{Ann}_H Q/B = I$

(4) $K \otimes_R H/I \cong Q/B$ *and* $\mathrm{Hom}_R(K,Q/B) \cong H/I$.

(5) *Let* $\mathcal{D}$ *be the divisible submodule of* B *and* Q(I) *the full ring of quotients of* H/I. *Then* $Q/\mathcal{D}$ *is the full ring of quotients of* $B/\mathcal{D}$, $Q/\mathcal{D} \subset Q(I)$, *and* $(H/I) \cap Q/\mathcal{D} = B/\mathcal{D}$. *Furthermore,* H/I *is the completion of* $B/\mathcal{D}$.

Proof. Let B/R be a proper divisible submodule of K, and let $I = \mathrm{Hom}_R(K,B/R)$. Since $H = \mathrm{Hom}_R(K,K)$ it is clear that I is a proper ideal of H. By Corollary 1.2, we have $K \otimes_R I = B/R$. It follows from Theorem 6.1 that B is a strongly unramified ring extension of R. By Theorem 6.4, we have an exact sequence:

$$0 \to I \to H \xrightarrow{\varphi} H(B) \to 0$$

and thus H/I is isomorphic as a ring to H(B). Since H(B) is a torsion-free R-module, we see that I is an unmixed ideal of rank 0 in H by Lemma 6.5.

By Corollary 2.5, we have an exact sequence:

$$0 \to \mathcal{D} \to B \to H(B) \to \mathrm{Ext}^1_B(Q,B) \to 0$$

where $\mathcal{D}$ is the divisible submodule of B. It follows from this that $K \otimes_R B \cong K \otimes_R H(B)$. Now $K \otimes_R B \cong Q/B$ by Theorem 1.1, since Q/B is a torsion R-module. Hence we have $Q/B \cong K \otimes_R H(B) \cong K \otimes_R H/I$. Since H/I is complete in the B-topology, it is complete in the R-topology by Theorem 3.2. Thus by Theorem 2.2 we have
$H/I \cong \mathrm{Hom}_R(K,K \otimes_R H/I) \cong \mathrm{Hom}_R(K,Q/B)$.

Because $H/I \cong \mathrm{Hom}_R(K,Q/B)$, we have $\mathrm{Ann}_H(Q/B) \subset I$. On the other hand we have $Q/B \cong K \otimes_R H/I$, and thus $I \subset \mathrm{Ann}_H(Q/B)$. Therefore, $\mathrm{Ann}_H(Q/B) = I$.

Let $\overline{B} = B/\mathcal{P}$ and $\overline{Q} = Q/\mathcal{P}$. Now $\mathcal{P}$ is an ideal of Q, and Q is a semi-local Noetherian ring of Krull dimension 0. Thus every regular element of $\overline{Q}$ is invertible in $\overline{Q}$. Hence by Theorem 2.11, $\overline{Q}$ is the full ring of quotients of $\overline{B}$. We also have by Theorem 2.11 that H(B) is the completion of $\overline{B}$ in the $\overline{B}$-topology. Since $H(B) \cong H/I$, we have an imbedding $\overline{B} \subset H/I \subset Q(I)$. This gives rise to an imbedding $\overline{Q} \subset Q(I)$ under which we have $\overline{B} \subset (H/I) \cap \overline{Q} = B_1$. Now $B_1/\overline{B}$ is a torsion-free $\overline{B}$-module because it is contained in $H(\overline{B})/\overline{B} = (H/I)/\overline{B}$. On the other hand $B_1/\overline{B}$ is a torsion $\overline{B}$-module because it is contained in $\overline{Q}/\overline{B}$. Thus $B_1/\overline{B} = 0$, and hence $\overline{B} = (H/I) \cap \overline{Q}$.

The only thing remaining to be proved is that if we start with an unmixed ideal I of rank 0 in H, then $I = \mathrm{Hom}_R(K, K \otimes_R I)$. Let $J = \mathrm{Hom}_R(K, K \otimes_R I)$; then by Theorem 2.2, $I \subset J$ and J/I is a torsion-free and divisible R-module. Hence $(HM)(J/I) = J/I$. But J is an ideal of H, and thus J/I is a finitely generated H-module. Since H is a local ring with **maximal** ideal HM, we see by the **Nakayama** Lemma that $J/I = 0$, and hence $J = I$.

<u>Corollary 6.7</u>. <u>Let</u> $I_1, \ldots, I_k$ <u>be a finite set of unmixed ideals of rank</u> 0 <u>in</u> H. <u>Then</u> $K \otimes_R (\bigcap_j I_j) = h(\bigcap_j (K \otimes_R I_j))$.

<u>Proof</u>. We have $\mathrm{Hom}_R(K, h(\bigcap_j (K \otimes_R I_j)) = \mathrm{Hom}_R(K, \bigcap_j (K \otimes_R I_J))$ $= \bigcap_j \mathrm{Hom}_R(K, K \otimes_R I_j)$. By Theorem 6.6 we see that $\bigcap_j \mathrm{Hom}_R(K, K \otimes_R I_j) = \bigcap_j I_j$; and hence by the same theorem, $K \otimes_R (\bigcap_j I_j) = h(\bigcap_j (K \otimes_R I_j))$.

<u>Remarks</u>. Since H is a **compatible ring extension of** R, an H-module is torsion, or divisible, or complete over H if and only if the same is true over R. By Theorem 3.1, $H \otimes_R Q$ is the full ring of quotients of H, and since $H \otimes_R K \cong K$ by Theorem 2.7, we see that $(H \otimes_R Q)/H \cong K$. By Theorem 2.10 there is a one-to-one correspondence between the set of rings B between R and Q and the set of rings Ω between H and $H \otimes_R Q$ given by $\Omega = H \otimes_R B$ and $B = \Omega \cap Q$. We have

$B/R \cong (H \otimes_R B)/H$ and $Q/B \cong (H \otimes_R Q)/(H \otimes_R B)$. It is also true that $B \subset H \otimes_R B$ and that $H \otimes_R B = HB = H + B$. Furthermore, $(H \otimes_R B)/B$ is torsion-free and divisible.

Theorem 6.8. (1) There is a one-to-one correspondence between the set of ideals $\mathcal{Q}$ of $H \otimes_R Q$ and set of the unmixed ideals I of rank 0 in H given by $\mathcal{Q} \cap H = I$ and $QI = \mathcal{Q}$.

(2) There is a one-to-one correspondence between the set of strongly unramified ring extensions Ω of H in $H \otimes_R Q$ and the set of ideals $\mathcal{Q}$ of $H \otimes_R Q$ given by $\Omega = H + \mathcal{Q}$ and $\mathcal{Q} = d(\Omega)$, the divisible submodule of Ω.

(3) There is a one-to-one correspondence between the set of ring extensions B of R in Q and the set of ring extensions Ω of H in $H \otimes_R Q$ given by $\Omega = H \otimes_R B$ and $B = \Omega \cap Q$. B is strongly unramified over R if and only if $H \otimes_R B$ is strongly unramified over H.

(4) If B and $\Omega = H \otimes_R B$ are corresponding strongly unramified ring extensions of R in Q and H in $H \otimes_R Q$, respectively, then $d(\Omega) \cap H = I$ is the unmixed ideal of rank 0 in H corresponding to B by Theorem 6.6. Furthermore $\Omega/d(\Omega) \cong H/I$ is the completion of both B and Ω in their respective topologies.

Proof. (1) H is a 1-dimensional, local, Cohen-Macaulay ring. Let $P_1,\ldots,P_n$ be the rank 0 prime ideals of H and $\mathcal{S} = H - \bigcup_{i=1}^{n} P_i$. Then $\mathcal{S}$ is the set of regular elements of H, and hence $H_{\mathcal{S}} = H \otimes_R Q$. There is clearly a one-to-one correspondence between the set of ideals $\mathcal{Q}$ of $H_{\mathcal{S}}$ and the set of unmixed ideals I of rank 0 in H given by $\mathcal{Q} \cap H = I$ and $I_{\mathcal{S}} = \mathcal{Q}$. But $I_{\mathcal{S}} = H_{\mathcal{S}}I = QHI = QI$.

(2) Let Ω be a strongly unramified ring extension of H in $H \otimes_R Q$, and let $\mathcal{Q} = d(\Omega)$. Then $\mathcal{Q}$ is a torsion-free and divisible H-module, and hence is an $H \otimes_R Q$-module; that is $\mathcal{Q}$ is an ideal of $H \otimes_R Q$. Suppose that $\Omega \neq H + \mathcal{Q}$. Let $H_1 = (H + \mathcal{Q})/\mathcal{Q}$ and

$Q_1 = (H \otimes_R Q)/\mathcal{D}$. Then $H_1 \cong H/(\mathcal{D} \cap H)$ is a complete, 1-dimensional local, Cohen-Macaulay ring and Q_1 is its full ring of quotients. Let $\Omega_1 = \Omega/\mathcal{D}$; then $H_1 \subset \Omega_1 \subset Q_1$ and Ω_1 is a reduced H_1-module. We see that $\Omega_1/H_1 \cong \Omega/(H + \mathcal{D})$ is a homomorphic image of Ω/H, and thus is a nonzero divisible H_1-module. Hence $\mathrm{Hom}_{H_1}(Q_1,\Omega_1/H_1) \neq 0$. However, we have an exact sequence:

$$\mathrm{Hom}_{H_1}(Q_1,\Omega_1) \to \mathrm{Hom}_{H_1}(Q_1,\Omega_1/H_1) \to \mathrm{Ext}^1_{H_1}(Q_1,H_1)$$

The end terms of this sequence are 0, since Ω_1 is reduced and H_1 is a complete ring. Therefore $\mathrm{Hom}_{H_1}(Q_1,\Omega_1/H_1) = 0$. This contradiction shows that $\Omega = H + \mathcal{D}$.

On the other hand if $\mathcal{D}$ is an ideal of $H \otimes_R Q$, then $\mathcal{D}$ is a divisible H-module and $(H + \mathcal{D})/\mathcal{D} \cong H/(H \cap \mathcal{D})$ is reduced. Thus $\mathcal{D}$ is the divisible submodule of $H + \mathcal{D}$. Since $(H + \mathcal{D})/H \cong \mathcal{D}/(H \cap \mathcal{D})$ is a divisible H-module, $H + \mathcal{D}$ is a strongly unramified extension of H in $H \otimes_R Q$.

(3) This statement has been proved in Theorem 2.10.

(4) Let B be a strongly unramified ring extension of R in Q and let $\Omega = H \otimes_R B$, and $I = d(\Omega) \cap H$. By (2) we have $\Omega = H + d(\Omega)$, and hence $\Omega/H \cong d(\Omega)/I$. Since $d(\Omega)$ is a torsion-free and divisible R-module, we see that $K \otimes_R I \cong \mathrm{Tor}_1^R(K,d(\Omega)/I) \cong t(d(\Omega)/I) = d(\Omega)/I$. By (3) we have $B/R \cong \Omega/H \cong d(\Omega)/I$, and therefore $K \otimes_R I \cong B/R$. It follows from Theorem 6.6 that I is the unmixed ideal of rank 0 in H corresponding to B. By Theorem 6.6, H/I is the completion of B. Now $\Omega/d(\Omega) = (H + d(\Omega))/d(\Omega) \cong H/I$, and hence H/I is the completion of Ω.

Remarks. Theorem 6.8 shows that every proper divisible submodule of K is of the form QI/I, where I is an unmixed ideal of rank 0 in H.

The next theorem is a generalization of the Theorem of Krull-Akiyuki. It shows that some of the strongly unramified extensions of R in Q may not be Noetherian. A different proof of this theorem may be found in [4, Prop. 6.4.13]. I am grateful to Mrs. Judy Sally for drawing it to my attention.

Theorem 6.9. Let R be a 1-dimensional, local, Cohen-Macaulay ring. Then every ring between R and Q is Noetherian if and only if R has no nonzero nilpotent elements.

Proof. Suppose that R has a nonzero nilpotent element c. Without loss of generality we can assume that $c^2 = 0$. Let b be a regular element of M, the maximal ideal of R, and let $S_n = R + R\,c/b^n$ for all $n > 0$. Then S_n is a ring, and $S_n \subset S_{n+1}$. We let S be the union of the S_n and I the ideal of S generated by the elements c/b^n. If I is a finitely generated ideal of S, then there exists an integer n and an element $s \in S$ such that $c/b^{n+1} = sc/b^n$. Now there is an integral $k > 0$ and elements $r, t \in R$ such that $s = r + tc/b^k$. It follows that $c/b^{n+1} = rc/b^n$. Therefore, $c(1-rb) = 0$, and since $1 - rb$ is a unit in R, we see that $c = 0$. This contradiction proves that S is not a Noetherian ring.

Conversely, suppose that R has no nilpotent elements $\neq 0$. Then $0 = p_1 \cap \ldots \cap p_t$, where $p_1, \ldots, p_t$ are the rank 0 prime ideals of R. It follows that R_{p_i} is a field for all $i = 1, \ldots, n$. Since $Q \cong R_{p_1} \oplus \ldots \oplus R_{p_t}$, Q is a direct sum of fields. Thus Q is a semi-simple ring.

Let B be a ring such that $R \subset B \subsetneq Q$. If $B'/R = h(B/R)$, then B' is a strongly unramified ring extension of R. By Theorem 5.1, B/B' is a finitely generated R-module. Thus without loss of generality, we can assume that B is a strongly unramified extension of R.

Let $\mathcal{D}$ be the divisible submodule of B. Then $\mathcal{D}$ is torsion-free and divisible, and hence $\mathcal{D}$ is a Q-module; that is, $\mathcal{D}$ is an ideal of Q. Since Q is a semi-simple ring, there is an idempotent element $e \in \mathcal{D}$ such that $Qe = \mathcal{D} = \mathcal{D}e$. It follows that $\mathcal{D} = Be$, and hence $\mathcal{D}$ is a principal ideal of B.

If we let $\overline{R} = R/(R \cap \mathcal{D})$, $\overline{B} = B/\mathcal{D}$, and $\overline{Q} = Q/\mathcal{D}$, then $\overline{B}$ is a ring extension of $\overline{R}$ in $\overline{Q}$. Furthermore $\overline{R}$ has the same properties as R and $\overline{Q}$ is the full ring of quotients of $\overline{R}$. Since θ is a finitely generated ideal of B, it follows that B is Noetherian if and only if $\overline{B}$ is Noetherian. Thus without loss of generality we may assume that B is a reduced R-module.

We proceed by induction on L(K). If S/R is a proper, nonzero, divisible submodule of B/R, then S is a strongly unramified ring extension of R. Since $S \subset B$, S is a reduced R-module. Thus by Theorem 3.6 (6), S is a quasi-local ring. Of course S has no nilpotent elements $\neq 0$. Because $L(Q/S) < L(K)$, it follows by induction that if we have proved that S is a Noetherian ring, then B is also a Noetherian ring. Thus without loss of generality we can assume that B/R has no proper, nonzero, divisible submodules. Hence by Theorem 5.1, every proper submodule of B/R is a finitely generated R-module.

Let J be an ideal of B and $I = J \cap R$. Suppose that J is not a finitely generated ideal of B. If J is a regular ideal of B, then $J = BI$ by Theorem 3.6 (2), and hence J is finitely generated over B. Thus J is not a regular ideal of B, and hence I is contained in the intersection of some of the prime ideals of rank 0 of R. Therefore there exists an element $c \in R$ that is not contained in any rank 0 prime ideal of R that contains I and such that $cI = 0$. $I + Rc$ is not contained in any rank 0 prime ideal of R; and hence there is a regular element $b \in I + Rc$. Thus $b = a + rc$ where $a \in I$ and $r \in R$.

Now Jb is not a finitely generated R-module. Hence $B = Jb + R$ because every proper submodule of B/R is finitely generated. It follows that $J = Jb + I$. Since $Ic = 0$, we see that $Jc = 0$, and hence $Jb = Ja \subset JI$. Therefore $J = BI$, and hence J is a finitely generated ideal of B. This contradiction proves that B is a Noetherian ring.

CHAPTER VII

THE CLOSED COMPONENTS OF R

Throughout this chapter R will be a 1-dimensional local Cohen-Macaulay ring.

Definitions. R is said to be analytically irreducible if its completion H is an integral domain. Of course R is then also an integral domain. We recall that we have defined R to be a closed domain if R is an integral domain and $J \cap R \neq 0$ for every nonzero ideal J of H.

Theorem 7.1. The following statements are equivalent:

(1) R is a closed domain.

(2) R is analytically irreducible.

(3) Every nonzero R-endomorphism of K is an epimorphism.

(4) K is a simple divisible module.

(5) E is a simple divisible module.

(6) Q is a field and the integral closure V of R in Q is a discrete valuation ring that is a finitely generated R-module.

Proof. (1) ⟹ (2). Let P be a prime ideal of rank 0 in H. Then $P \cap R$ is a prime ideal of R and $P \cap R \neq M$. Thus $P \cap R = 0$, and hence $P = 0$. Therefore, H is an integral domain.

(2) ⟹ (3). Let f be a nonzero element of $H = \mathrm{Hom}_R(K,K)$. Suppose that f is not an epimorphism on K. Then $L(f(K)) < L(K)$, and hence by Theorem 5.11, $L(\mathrm{Ker}\, f) \neq 0$; that is Ker f is not reduced. Hence by Corollary 1.3 there is a nonzero element $g \in H$ such that $g(K) \subset \mathrm{Ker}\, f$. We then have $fg = 0$, contradicting the fact that H is an integral domain. Thus every element of H is an epimorphism on K.

(3) ⟹ (4). Let B be a proper, nonzero R-submodule of K. If B is not reduced, then by Corollary 1.3, there is an R-endomorphism of K that is not an epimorphism. Hence B is reduced, and thus K is

a simple divisible module.

(4) $\Longrightarrow$ (6). Suppose that Q is not a field. Then Q has a proper nonzero ideal $\mathfrak{P}$. Since $\mathfrak{P}$ is a divisible R-module, $\mathfrak{P}$ is not a finitely generated R-module by the Nakayama Lemma. It follows that $\mathfrak{P} + R = Q$. But then $Q/\mathfrak{P} \cong R/(\mathfrak{P} \cap R)$, and this is a contradiction since $Q/\mathfrak{P}$ has Krull dimension 0 and $R/(\mathfrak{P} \cap R)$ has Krull dimension 1. Therefore, Q is a field.

Let V be the integral closure of R in Q; then V/R is a proper submodule of K, and hence V is a finitely generated R-module by Theorem 5.1. Since every localization of V with respect to a maximal ideal of V is a finitely generated R-module, we see that V is a local ring. Therefore, V is a Noetherian, integrally closed, local domain of Krull dimension 1 by Corollary 6.2 and hence V is a discrete valuation ring.

(6) $\Longrightarrow$ (2). By Theorem 6.4, H is a subring of H(V), and H(V) is an integral domain.

(2) $\Longrightarrow$ (1). Let J be an ideal of H that is maximal with respect to the property $J \cap R = 0$. It is easily seen that J is a prime ideal of H, and since $HM \cap R = M$, we have $J = 0$.

(1) $\Longrightarrow$ (5). Let D be a nonzero divisible R-submodule of E, and let $J = \mathrm{Ann}_H D$. Then J is an ideal of H and $J \cap R = 0$. Hence we have $J = 0$. By Theorem 4.5, we have $D = \mathrm{Ann}_E J = \mathrm{Ann}_E 0 = E$, and thus E is a simple divisible R-module.

(5) $\Longrightarrow$ (2). Since E is a simple divisible R-module, every R-endomorphism of E is an epimorphism. Thus $\mathrm{Hom}_R(E,E)$ has no zero divisors. But $H \cong \mathrm{Hom}_R(E,E)$ by Theorem 4.5, and thus H is an integral domain.

<u>Remarks</u>. We note that it follows from Corollary 5.7 and Theorem 7.1 that R is analytically irreducible if and only if every proper submodule of K (or of E) is a finitely generated R-module--in fact,

if and only if it has finite length.

Definition. We shall call a **maximal** strongly unramified extension of R in Q *a closed component of R*. By Theorem 6.1, A is a closed component of R if and only if A/R is a maximal proper divisible submodule of K. By Theorem 5.5, K has ACC on divisible submodules, and thus closed components of R exist.

Theorem 7.2. Let A be a strongly unramified ring extension of R in Q. Then the following statements are equivalent.

(1) A is a closed component of R.

(2) Q/A is a simple divisible A-module.

(3) H(A), the completion of A in the A-topology, is an integral domain.

(4) If $\mathcal{D}$ is the divisible submodule of A, then A/$\mathcal{D}$ is an analytically irreducible, Noetherian, local domain of Krull dimension 1.

Proof. (1) $\Longleftrightarrow$ (2). By Theorem 3.2, an A-module is a torsion, divisible A-module if and only if it is a torsion, divisible R-module. Furthermore, by Theorem 3.6 (7), every R-module between A and Q is also an A-module. Thus A is a closed component of R if and only if Q/A is a simple divisible A-module.

(2) $\Longrightarrow$ (3). Since Q/A is a simple divisible A-module, every element of $H(A) = \mathrm{Hom}_A(Q/A, Q/A)$ is an epimorphism. Therefore H(A) has no zero divisors. Since H(A) is a commutative ring by Corollary 2.5, H(A) is an integral domain.

(3) $\Longrightarrow$ (4). If $\mathcal{D}$ is the divisible submodule of A, then $\mathcal{D}$ is the kernel of the ring homomorphism $A \to H(A)$, and thus $\mathcal{D}$ is a prime ideal of A. $\mathcal{D}$ is a Q-module and hence $\mathcal{D}$ is an ideal of Q. Let $\overline{R} = R/(\mathcal{D} \cap R)$, $\overline{A} = A/\mathcal{D}$, and $\overline{Q} = Q/\mathcal{D}$. Then $\overline{R}$ is a Noetherian, local domain of Krull dimension 1 and $\overline{Q}$ is the quotient field of $\overline{R}$.

Since $\overline{A}/\overline{R} \cong A/(R + \mathcal{D})$ is a homomorphism image of A/R, it follows that $\overline{A}/\overline{R}$ is a divisible $\overline{R}$-module. Therefore, $\overline{A}$ is a strongly unramified ring extension of $\overline{R}$. By Theorem 6.1, $\overline{A}$ is a Noetherian local domain of Krull dimension 1. Since $H(\overline{A}) = H(A)$ is an integral domain, $\overline{A}$ is analytically irreducible.

(4) ==> (2). We shall use the notation of the preceding paragraph. By Theorem 7.1, $\overline{Q}/\overline{A}$ is a simple divisible $\overline{A}$-module. Since $\overline{Q}/\overline{A} \cong Q/A$, we see that Q/A is a simple divisible A-module.

Remarks. Let A be a closed component of R, and let $\mathcal{D}$ be the divisible submodule of A. Since $\overline{A} = A/\mathcal{D}$ is an integral domain and $\overline{Q} = A/\mathcal{D}$ is the full ring of quotients of $\overline{A}$, we see that $\overline{Q}$ is a field and thus $\mathcal{D}$ is a maximal ideal of Q. Since the sum of two maximal ideals of Q is equal to Q, we observe that $\mathcal{D}$ is the unique maximal ideal of Q contained in A.

Let $\overline{R} = R/(\mathcal{D} \cap R)$. Then $\overline{R}$ is a Noetherian local domain of Krull dimension 1, and $\overline{Q}$ is the quotient field of $\overline{R}$. Since $\overline{Q}/\overline{A} \cong Q/A$, we see that $\overline{Q}/\overline{A}$ is a simple A-module. Thus $\overline{A}$ is a closed component of $\overline{R}$.

Because $Q/A \cong (H \otimes_R Q)/(H \otimes_R A)$ by Theorem 2.10, we see that $H \otimes_R A$ is a closed component of H. Furthermore, every closed component of H arises in this way. By Theorem 6.8, $H \otimes_R A = H + L$, where L is an ideal of $H \otimes_R Q$ and L is the divisible submodule of $H \otimes_R A$. But then by the previous paragraph L is a maximal ideal of $H \otimes_R Q$. Because $\mathcal{D} \subset L$, and $\mathcal{D}$ is a maximal ideal of Q, it follows that $\mathcal{D} = Q \cap L$. By Theorem 6.8 the completions of A, $A/\mathcal{D}$, and $H \otimes_R A$ are all equal to $H/(H \cap L)$. We shall enlarge upon these remarks in the next theorem.

Definition. If P is a prime ideal of rank 0 in H, then H/P is called an analytic component of R. H/P is a complete, Noetherian, local domain of Krull dimension 1.

If B is a ring, we recall that we denote its completion in the B-topology by H(B).

Theorem 7.3. There is a one-to-one correspondence between the set of closed components A of R and the set of prime ideals P of rank 0 in H; and hence there is a one-to-one correspondence between the set of closed components of R and the set of analytic components of R. The correspondence is given by:

$$K \otimes_R P = A/R \text{ and } \mathrm{Hom}_R(K, A/R) = P.$$

We then have:

(1) $P = \mathrm{Ann}_H Q/A$

(2) $H(A) = H/P$

(3) *If* $\mathcal{D}$ *is the divisible submodule of* A, *and* L *the divisible submodule of* $H \otimes_R A$, *then* $L \cap H = P$, $H \otimes_R A = H + L$, *and* $(H/P) \cap Q/\mathcal{D} = A/\mathcal{D}$. *Furthermore,* H/P *is the completion of* $A/\mathcal{D}$.

Proof. Let A be a closed component of R. Then by Theorem 7.2, H(A) is an integral domain. If we let $P = \mathrm{Hom}_R(K, A/R)$, then by Theorem 6.6, $H(A) \cong H/P$. Hence P is a prime ideal of H, necessarily of rank 0. Conversely, if P is a prime ideal of rank 0 in H, then by Theorem 6.6, $K \otimes_R P = A/R$, where A is a strongly unramified extension of R in Q. By the order-preserving nature of the one-to-one correspondence of Theorem 6.6, A/R is a maximal proper divisible submodule of K, and thus A is a closed component of R. The remaining statements of the theorem follow from Theorem 6.6 and the Remarks preceding the theorem.

Definition. If P is a prime ideal of rank 0 in H and A is the corresponding closed component of R in Q, we shall denote this by writing $P = P_A$ and $A = A_P$.

Theorem 7.4. If P is a prime ideal of rank 0 in H, then $P = \mathrm{Ann}_H(\mathrm{Ann}_K P)$.

Proof. It is clear that $P \subset \mathrm{Ann}_H(\mathrm{Ann}_K P) = I$. Since P belongs to 0 in H, there is a nonzero element $f \in H$ such that $Pf = 0$. Thus $f(K) \subset \mathrm{Ann}_K P$, and hence $\mathrm{Ann}_K P$ is not reduced. It follows that I contains no regular elements of H, and hence I is contained in a prime ideal P' of H of rank 0. Since $P \subset P'$, we see that $P = P'$, and hence $P = \mathrm{Ann}_H(\mathrm{Ann}_K P)$.

Definition. A ring W is called a valuation ring if it is an integral domain such that if x is a nonzero element of the quotient field of W, then either x or 1/x is in W. A valuation ring is quasi-local. A valuation ring is called a discrete valuation ring if it is a Noetherian ring. A discrete valuation ring is a Noetherian local domain of Krull dimension 1, and every ideal is a principal ideal.

We shall say that a commutative ring V is a pseudo valuation ring if module its divisible submodule it is a valuation ring. The divisible submodule of a pseudo valuation ring is thus a prime ideal. If R is a 1-dimensional, local, Cohen-Macaulay ring, then the pseudo valuation rings between R and Q will be called the pseudo valuation rings belonging to R.

Theorem 7.5. There is a one-to-one correspondence between the set of closed components A of R and the set of pseudo valuation rings V belonging to R given by $h(V/R) = A/R$ and (Integral Closure of A in Q) = V.

Proof. Let A be a closed component of R, and let $\mathcal{D}$ be the divisible submodule of A. Then $\mathcal{D}$ is a maximal ideal of Q and $Q/\mathcal{D}$ is the quotient field of $A/\mathcal{D}$. By Theorem 7.2, $A/\mathcal{D}$ is an analytically irreducible, Noetherian, local domain of Krull dimension 1. Let $V/\mathcal{D}$ be the integral closure of $A/\mathcal{D}$ in $Q/\mathcal{D}$. Since $\mathcal{D}$ is an ideal of Q, it is easy to see that V is the integral closure of A in

Q. By Theorem 7.1, $V/\mathcal{P}$ is a discrete valuation ring. Therefore V is a pseudo valuation ring belonging to R.

Conversely, let V be a pseudo valuation ring belonging to R, and let $A/R = h(V/R)$. Then A is a strongly unramified ring extension of R in Q by Theorem 6.1. Since V/A is a reduced R-module, it is a finitely generated torsion R-module by Theorem 5.1. Hence V/A has bounded order. Thus by Theorem 2.9 we have $H(A) \subset H(V)$. Let $\mathcal{P}$ be the divisible submodule of V. Then $H(V) = H(V/\mathcal{P})$ by Theorem 2.11. Since $V/\mathcal{P}$ is a valuation ring by definition, $H(V/\mathcal{P})$ is also a valuation ring. Because $H(A) \subset H(V/\mathcal{P})$, we conclude that H(A) is an integral domain. Therefore, A is a closed component of R by Theorem 7.2.

Corollary 7.6. Let V be a pseudo valuation ring belonging to R, and A the closed component of R corresponding to V. Then:

(1) V/A is a torsion R-module of finite length.

(2) If $\mathcal{P}$ is the divisible submodule of V, then $\mathcal{P}$ is the divisible submodule of A.

(3) $\mathcal{P}$ is a maximal ideal of Q.

(4) $V/\mathcal{P}$ is a discrete valuation ring belonging to $R/(\mathcal{P} \cap R)$.

(5) $V/\mathcal{P}$ is the integral closure of $A/\mathcal{P}$, and $A/\mathcal{P}$ is a closed component of $R/(\mathcal{P} \cap R)$.

Proof. If $\mathcal{P}'$ is the divisible submodule of A, then $\mathcal{P}' \subset \mathcal{P}$; and both $\mathcal{P}'$ and $\mathcal{P}$ are ideals of Q. Since $\mathcal{P}'$ is a maximal ideal of Q, we see that $\mathcal{P}' = \mathcal{P}$. Now $R/(\mathcal{P} \cap R)$ is a Noetherian local domain of Krull dimension 1 and quotient field $Q/\mathcal{P}$. Thus $V/\mathcal{P}$ is a Noetherian ring by Corollary 6.2, and hence $V/\mathcal{P}$ is a discrete valuation ring.

The other statements of the corollary were proved in the course of proving Theorem 7.5.

Theorem 7.7. The integral closure of R in Q is equal to the

<u>intersection of all of the pseudo valuation rings belonging to</u> R.

<u>Proof</u>. If an element of Q is integral over R, then it is integral over every closed component of R. Hence by Theorem 7.5 the integral closure of R is contained in every pseudo valuation ring belonging to R.

Let $\mathcal{P}_1, \ldots, \mathcal{P}_t$ be the set of maximal ideals of Q, and for each $i = 1, \ldots, t$ let W_i be the intersection of those pseudo valuation rings of R that contain $\mathcal{P}_i$. Then $W_i/\mathcal{P}$ is the intersection of all of the valuation rings belonging to $R/(R \cap \mathcal{P}_i)$, and hence $W_i/\mathcal{P}_i$ is the integral closure of $R/(R \cap \mathcal{P}_i)$ in $Q/\mathcal{P}_i$.

Suppose that x is an element of the intersection of all of the pseudo valuation rings belonging to R. Then $x \in W_1$, and hence there is an integer $n_1 > 0$ and a polynomial $f_1(X)$ of degree $< n_1$ with coefficients in R such that $x^{n_1} - f_1(x) \in \mathcal{P}_1$. Let $y_1 = x^{n_1} - f_1(x)$; then $y_1 \in W_2$, and hence there is an integer $n_2 > 0$ and a polynomial $f_2(X)$ of degree $< n_2$ such that $y_1^{n_2} - f_2(y_1) \in \mathcal{P}_1 \cap \mathcal{P}_2$. Continuing in this way we see that there is a polynomial $g(X)$ of degree $< n = n_1 n_2 \ldots n_t$ such that $x^n - g(x) \in \mathcal{P}_1 \cap \ldots \cap \mathcal{P}_t$. But every element of $\mathcal{P}_1 \cap \ldots \cap \mathcal{P}_t$ is nilpotent, and hence there exists an integer $k > 0$ such that $[x^n - g(x)]^k = 0$. This equation shows that x is integral over R.

<u>Theorem 7.8</u>. <u>Let</u> $A_1, \ldots, A_n$ <u>be the closed components of</u> R <u>and</u> $V_1, \ldots, V_n$ <u>the pseudo valuation rings belonging to</u> R. <u>Let</u> $T = \bigcap_i A_i$ <u>and</u> $S = \bigcap_i V_i$. <u>Then</u> S <u>is the integral closure of</u> R <u>in</u> Q, <u>and</u>:

(1) $Q/T \cong \sum_{k=1}^{n} \oplus\, Q/A_k$ <u>and</u> $A_k + \bigcap_{m \neq k} A_m = Q$.

(2) $Q/S \cong \sum_{k=1}^{n} \oplus\, Q/V_t$ <u>and</u> $V_k + \bigcap_{m \neq k} V_m = Q$.

(3) S/T <u>is an</u> R-<u>module of finite length, and hence</u> S <u>is a finitely generated</u> T-<u>module</u>.

<u>Proof</u>. (1) Let $\mathcal{P}_1, \ldots, \mathcal{P}_t$ be the maximal ideals of Q and

for each $i = 1,\ldots,t$ let B_i be the intersection of those closed components of R that contain $\mathcal{P}_i$. Since $\mathcal{P}_1 + (\mathcal{P}_2 \cap \ldots \cap \mathcal{P}_t) = Q$, we see that $B_1 + (B_2 \cap \ldots \cap B_t) = Q$. Since $T = \bigcap_{i=1}^{t} B_i$, we see that $Q/T = (B_1/T) \oplus (B_2 \cap \ldots \cap B_t)/T \cong Q/(B_2 \cap \ldots \cap B_t) \oplus Q/B_1$. By induction on t we have $Q/(B_2 \cap \ldots \cap B_t) \cong Q/B_2 \oplus \ldots \oplus Q/B_t$. Thus $Q/T \cong Q/B_1 \oplus \ldots \oplus Q/B_t$.

Let $\overline{R}_i = R/(R \cap \mathcal{P}_i)$, $\overline{Q}_i = Q/\mathcal{P}_i$ and $\overline{B}_i = B_i/\mathcal{P}_i$. Let $A_{1i},\ldots,A_{pi}$ be the closed components of R that contain $\mathcal{P}_i$ and let $\overline{A}_{ji} = A_{ji}/\mathcal{P}_i$. Then $\overline{R}_i$ is a Noetherian local domain of Krull dimension 1, and $\overline{A}_{1i},\ldots,\overline{A}_{pi}$ are the closed components of $\overline{R}_i$. Since $\overline{A}_{ji} + \overline{A}_{mi} = \overline{Q}_i$ for $j \neq m$, we can apply Theorem 6.3 and obtain $\overline{Q}_i/\overline{B}_i \cong \overline{Q}_i/\overline{A}_{1i} \oplus \ldots \oplus \overline{Q}_i/\overline{A}_{pi}$. But $\overline{Q}_i/\overline{B}_i \cong Q/B_i$ and $\overline{Q}_i/\overline{A}_{ji} \cong Q/A_{ji}$. Therefore $Q/T \cong \sum_{i=1}^{t} \oplus Q/B_i \cong \sum_{i,j} \oplus Q/A_{ji} = \sum_{k=1}^{n} \oplus Q/A_k$.

Let $C_k = \bigcap_{m \neq k} A_m$; then we have an exact sequence

$$0 \to C_k/T \to Q/T \to Q/C_k \to 0.$$

Then by what we have already proved, $Q/T \cong \sum_{k=1}^{n} \oplus Q/A_k$ and $Q/C_k \cong \sum_{m \neq k} \oplus Q/A_m$, and hence $L(Q/T) = n$ and $L(Q/C_k) = n - 1$. Thus $L(C_k/T) = 1$ and hence C_k/T is not h-reduced. But $C_k/T \cong (C_k + A_k)/A_k$, and since Q/A_k is a simple divisible R-module, we see that $C_k + A_k = Q$.

(2) Since $V_k \supset A_k$ for $k = 1,\ldots,n$, we see by (1) that $V_k + \bigcap_{m \neq k} V_m = Q$. Hence $Q/S \cong V_k/S \oplus (\bigcap_{m \neq k} V_m)/S \cong Q/(\bigcap_{m \neq k} V_m) \oplus Q/V_k$. By induction on n we have $Q/(\bigcap_{m \neq k} V_m) \cong \sum_{m \neq k} \oplus Q/V_m$, and hence $Q/S \cong \sum_{k=1}^{n} \oplus Q/V_t$.

(3) As we observed in Theorem 7.5, if V_k is the pseudo valuation ring of R corresponding to the closed component A_k of R, then V_k/A_k is an R-module of finite length. Hence $L(Q/V_k) = L(Q/A_k) = 1$.

Thus by (1) and (2) we have $L(Q/T) = n = L(Q/S)$, and hence $L(S/T) = 0$. Therefore, by Theorem 5.1, S/T is an R-module of finite length.

Remarks. Suppose that R is a Noetherian domain of Krull dimension 1 so that the pseudo valuation rings of R are in fact bonafide valuation rings. In this case if S is the integral closure of R in Q, then S is a semi-local principal ideal domain by Corollary 6.2. If $N_1,\ldots,N_n$ are the maximal ideals of S, then $S_{N_1},\ldots,S_{N_n}$ are exactly the valuation rings belonging to R. We then have $Q/S \cong Q/S_{N_1} \oplus \ldots \oplus Q/S_{N_n}$ by Theorem 4.1, and this fact could have been used to give a slightly different proof of Theorem 7.8.

The next theorem was proved in the case where R is an integral domain by Northcott [19, Th. 5].

Theorem 7.9. There is a one-to-one correspondence between the set of pseudo valuation rings belonging to R and the set of prime ideals of rank 0 in H.

Proof. This is an immediate consequence of putting together Theorems 7.3 and 7.5.

Remarks. Let V be a pseudo valuation ring of R and let A be the corresponding closed component of R. Then $H \otimes_R A$ is a closed component of H in $H \otimes_R Q$. To show that $H \otimes_R V$ is the corresponding pseudo valuation ring of H it is sufficient by Theorem 7.5 to show that $H \otimes_R V$ is the integral closure of $H \otimes_R A$ in $H \otimes_R Q$.

Now on the one hand $H \otimes_R V$ is an integral extension of $H \otimes_R A$ in $H \otimes_R Q$ because $(H \otimes_R V)/(H \otimes_R A) \cong V/A$ is a finitely generated H-module. On the other hand if Ω is the integral closure of $H \otimes_R A$ in $H \otimes_R Q$, and $B = \Omega \cap Q$, then by Theorem 2.10, $\Omega = H \otimes_R B$ and $B/A \cong (H \otimes_R B)/(H \otimes_R A)$. Hence by Corollary 7.6 applied to H, we see that B/A is an H-module of finite length. Therefore, B is integral over A and hence $B \subset V$. Thus $\Omega = H \otimes_R B \subset H \otimes_R V$, proving that

$H \otimes_R V$ is the integral closure of $H \otimes_R A$ in $H \otimes_R Q$.

Let $\mathcal{D}$ be the divisible submodule of V, and L the divisible submodule of $H \otimes_R V$. Then $\mathcal{D}$ and L are the divisible submodules of A and $H \otimes_R A$, respectively. Hence $H \otimes_R A = H + L$ and $\mathcal{D} = L \cap Q$. If we let $P = L \cap H$, then $H/P \cong (H \otimes_R A)/L$ is the completion of both A and $H \otimes_R A$, and hence P is the prime ideal of rank 0 in H corresponding to A by Theorem 7.3.

By Corollary 7.6, $(H \otimes_R V)/L$ is a discrete valuation ring, and it is the integral closure of $(H \otimes_R A)/L = H/P$ in its quotient field. Because $(H \otimes_R V)/L$ is finitely generated as an H/P-module, it is a complete discrete valuation ring. Furthermore, $(H \otimes_R V)/L \cap Q/\mathcal{D} = V/\mathcal{D}$, and thus $(H \otimes_R V)/L$ is the completion of $V/\mathcal{D}$.

All of these remarks go to show the nature of the one-to-one correspondence between the set of pseudo valuation rings belonging to R, and the set of complete discrete valuation rings which are the integral closures of the analytic components of R.

CHAPTER VIII
SIMPLE DIVISIBLE MODULES

Throughout this chapter R will be a 1-dimensional, local, Cohen-Macaulay ring.

Lemma 8.1. *If T_1 and T_2 are equivalent torsion R-modules, then $\mathrm{Ann}_H T_1 = \mathrm{Ann}_H T_2$.*

Proof. By definition, we have R-epimorphisms $f : T_1 \to T_2$ and $g : T_2 \to T_1$. By Theorem 2.7, T_1 and T_2 are H-modules and f and g are H-homomorphisms. It follows immediately from this that $\mathrm{Ann}_H T_1 = \mathrm{Ann}_H T_2$.

Definition. Let T be a torsion R-module, and [T] its equivalence class. Because of Lemma 8.1, we can unambiguously define $\mathrm{Ann}_H([T])$ to be $\mathrm{Ann}_H T$.

Theorem 8.2. *Let D be a simple divisible R-module, f an R-homomorphism of K onto D, C/D = Ker f, and A/R = h(Ker f). Then:*

(1) *A is a closed component of R and is uniquely determined by the equivalence class of D.*

(2) *D is isomorphic to Q/C, and C is isomorphic to a regular ideal of A.*

(3) *If I is a regular ideal of A, then Q/I is a simple divisible R-module equivalent to D. Conversely, if D' is any simple divisible R-module equivalent to D, then D' $\cong$ Q/I for some regular ideal I of A.*

(4) *D is equivalent to Q/A and $\mathrm{Ann}_H D = P_A$.*

Proof. By Theorem 5.11 we have L(A/R) = L(C/R) = L(K) - L(D) = L(K) - 1. Thus A is a closed component of R. By Theorem 3.6, C is an A-module; and by Theorem 5.1, C/A is a finitely generated torsion R-module. Therefore, C is a finitely generated A-module, and hence C is isomorphic to a regular ideal of A. Because C/R = Ker f, it follows that $D \cong K/(\mathrm{Ker}\ f) \cong Q/C$.

Now $Q/C \cong (Q/A)/(C/A)$, and hence D is equivalent to Q/A by Lemma 5.8. If I is a regular ideal of A, then A/I is a cyclic torsion R-module by Theorem 3.6. Therefore, Q/I is also equivalent to Q/A by Lemma 5.8.

By Lemma 8.1 we have $\mathrm{Ann}_H D = \mathrm{Ann}_H(Q/A)$, and by Theorem 7.3 we have $\mathrm{Ann}_H(Q/A) = P_A$. Therefore, $\mathrm{Ann}_H D = P_A$. Because A and P_A uniquely correspond, and because $\mathrm{Ann}_H D$ is an invariant of the equivalence class of D, we see that A is independent of the homomorphism f and of the representative of the equivalence class of D. By what we have already proved this shows that if D' is a simple divisible R-module equivalent to D, then D' is isomorphic to Q/I, where I is a regular ideal of A.

Corollary 8.3. (1) *There is a one-to-one correspondence between the set of equivalence classes* [D] *of simple divisible* R-*modules, and the set of closed components* A *of* R *in* Q *given by* $Q/A \sim D$ *and* $A/R = h(\mathrm{Ker}\ f)$, *where* f *is any* R-*homomorphism of* K *onto* D.

(2) *There is also a one-to-one correspondence between the set of classes* [D] *and the set of prime ideals* P *of rank* 0 *in* H *given by* $P = \mathrm{Ann}_H([D])$ *and* $D \sim K \otimes_R H/P$.

(3) P *and* [D] *correspond, and* A *and* [D] *correspond if and only if* $P = P_A$ *and* $Q/A \sim D$.

Proof. If A is a closed component of R in Q, then $L(A/R) = L(K) - 1$ and $L(Q/A) = L(K) - L(A/R)$. Therefore, $L(Q/A) = 1$; and hence Q/A is a simple divisible R-module. If P is a rank 0 prime ideal of H, then by Theorem 6.6 and 7.3 we have $K \otimes_R (H/P) \cong Q/A_P$. The remainder of the theorem follows from Theorem 8.2.

Corollary 8.4. *Let* $V_1, \ldots, V_n$ *be the pseudo valuation rings of* R. *Then* $Q/V_1, \ldots, Q/V_n$ *are a full set of representatives of the equivalence classes of simple divisible* R-*modules and* Q/V_i *is not* equivalent to Q/V_j if $i \neq j$.

Proof. Let A_i be the closed component of R corresponding to V_i by Theorem 7.5. By Corollary 7.6, V_i is a finitely generated A_i-module, and hence V_i is isomorphic to a regular ideal of A_i. The corollary now follows from Theorems 8.2 and Corollary 8.3.

Definitions. If D is a simple divisible R-module we shall let $G([D])$ denote the set of isomorphism classes of R-modules D' equivalent to D.

If A is any ring we shall let $G(A)$ denote the semi-group of isomorphism classes of regular ideals of A.

Theorem 8.5. Let D be a simple divisible R-module, let $P = \mathrm{Ann}_H(D)$, and let $\bar{H} = H/P$, the corresponding analytic component of R. Then there is a one-to-one correspondence between the sets $G([D])$ and $G(\bar{H})$ given by $\bar{I} \cong \mathrm{Hom}_R(K, D')$ and $D' \cong K \otimes_R \bar{I}$, where $\bar{I}$ is a nonzero ideal of $\bar{H}$ and D' is an R-module equivalent to D.

Proof. Let D' be an R-module equivalent to D. If A is the closed component of R corresponding to the equivalence class [D], then by Theorem 8.2 there is a regular ideal I of A such that $D' \cong K \otimes_R I$. Let $\bar{I} = \mathrm{Hom}_R(K, D') \cong \mathrm{Hom}_R(K, K \otimes_R I)$; then by Theorem 2.2, $\bar{I}$ is isomorphic to the completion of I in the R-topology, and hence by Theorem 3.2 to the completion of I in the A-topology. Now $\bar{H}$ is the completion of A by Theorem 7.3, and hence by Theorem 2.8, $\bar{I}$ is isomorphic to a nonzero ideal of $\bar{H}$. By Corollary 1.2 we have $K \otimes_R \bar{I} \cong K \otimes_R I \cong D'$.

On the other hand suppose that $\bar{I}$ is a nonzero ideal of $\bar{H}$. If $\mathcal{D}$ is the divisible submodule of A, then $A/\mathcal{D}$ is an analytically irreducible Noetherian local domain by Theorem 7.2; and $\bar{H}$ is its completion by Theorem 7.3. Hence $A/\mathcal{D}$ is a closed domain by Theorem 7.1, and thus $\bar{I} \cap A/\mathcal{D}$ is a nonzero ideal of $A/\mathcal{D}$. By Theorem 2.11 there is a regular ideal I of A such that $I/\mathcal{D} = \bar{I} \cap A/\mathcal{D}$. By Theorems

2.8 and 2.11 the completion of I is isomorphic to $\overline{I}$, and thus by Theorem 2.2 we have $\overline{I} \cong \mathrm{Hom}_R(K, K \otimes_R I)$. By Corollary 1.2, $K \otimes_R \overline{I} \cong K \otimes_R I$, and thus by Theorem 8.2, $K \otimes_R \overline{I}$ is equivalent to D.

Corollary 8.6. *Let D be a simple divisible R-module, and let A be the closed component of R corresponding to D. Then every R-module equivalent to D is isomorphic to D if and only if A is a pseudo valuation ring.*

Proof. A is a pseudo valuation ring if and only if its completion $\overline{H}$ is a complete discrete valuation ring. It is also clear that $\overline{H}$ is a complete discrete valuation ring if and only $G(\overline{H})$ consists of one element. Thus the corollary is an immediate consequence of Theorem 8.5.

Remarks. (1) Corollary 8.6 shows that modules can be equivalent without being isomorphic, and thus settles any lingering doubts that the whole concept of equivalence is just a vacuous extension of isomorphism. This, of course, could have been seen much earlier.

(2) Schur's Lemma states that the endomorphism ring of a simple module is a division ring. By analogy we should expect that the endomorphism ring of a simple divisible R-module should be something special. That this is indeed the case is shown by the following theorem.

Theorem 8.7. *Let D be a simple divisible R-module and $\overline{H}$ the corresponding analytic component of R. Then $\Gamma = \mathrm{Hom}_R(D, D)$ is a complete, Noetherian, local domain of Krull dimension 1. Γ is an extension ring of $\overline{H}$ in its quotient field and is a finitely generated $\overline{H}$-module. Conversely, if Γ is a ring with these properties then there is a simple divisible R-module D' equivalent to D such that* $\Gamma \cong \mathrm{Hom}_R(D', D')$.

Proof:

By Theorem 8.5, there is an ideal $\overline{I}$ of $\overline{H}$ such that $D \cong K \otimes_R \overline{I}$. Since $\overline{I}$ is a complete R-module, we have $\mathrm{Hom}_R(D,D) \cong \mathrm{Hom}_R(\overline{I},\overline{I})$ by Corollary 2.4. By Theorem 2.7, we have $\mathrm{Hom}_R(\overline{I},\overline{I}) \cong \mathrm{Hom}_H(\overline{I},\overline{I})$. Because $\overline{I}$ is an $\overline{H}$-module, we have $\mathrm{Hom}_H(\overline{I},\overline{I}) = \mathrm{Hom}_{\overline{H}}(\overline{I},\overline{I})$. Hence the statements about Γ follow from Corollary 6.2 and Theorem 7.1.

Conversely, let Γ be a domain with the properties defined in the theorem. Let $D' = K \otimes_R \Gamma$; since Γ is isomorphic to an ideal of $\overline{H}$, we see by Theorem 8.5 that D' is a simple divisible R-module equivalent to D. By the same arguments as in the preceding paragraph we have $\mathrm{Hom}_R(D',D') \cong \mathrm{Hom}_{\overline{H}}(\Gamma,\Gamma) \cong \Gamma$.

Theorem 8.8. *Let D_1 and D_2 be simple divisible R-modules. Then the following statements are equivalent:*

(1) *D_1 is equivalent to D_2.*

(2) $\mathrm{Hom}_R(D_1,D_2) \neq 0$.

(3) $\mathrm{Ann}_H D_1 = \mathrm{Ann}_H D_2$.

Proof. The equivalence of (1) and (2) is due to Lemma 5.8; and the equivalence of (1) and (3) is due to Corollary 8.3.

Corollary 8.9. *If D is a simple divisible R-module, then there is a submodule of K equivalent to D.*

Proof. If $P = \mathrm{Ann}_H(D)$, then we have seen in the course of the proof of Theorem 7.4 that $\mathrm{Ann}_K(P)$ is not reduced. Hence there is a simple divisible R-module D' contained in $\mathrm{Ann}_K(P)$. Since $P \subset \mathrm{Ann}_H D' = P'$ we see that $P = P'$. But then D and D' are equivalent by Corollary 8.3.

Corollary 8.10. *If D is a divisible R-module, then $\mathrm{Hom}_R(D,K) \neq 0$. Thus if B is a ring extension of R in Q, $B \neq Q$, then $\mathrm{Hom}_R(Q/B,K) \neq 0$ and $\mathrm{Hom}_R(B,H) \neq 0$.*

Proof. Since E is a universal injective R-module, we have $\mathrm{Hom}_R(D,E) \neq 0$. Let f be a nonzero element of $\mathrm{Hom}_R(D,E)$. Then Im f is an Artinian divisible R-module and $\mathrm{Hom}_R(\mathrm{Im}\ f,\ K) \subset \mathrm{Hom}_R(D,K)$. Thus we can assume that D is an Artinian divisible R-module. There is a submodule C of D such that D/C is a simple divisible R-module. Since $\mathrm{Hom}_R(D/C,K) \subset \mathrm{Hom}_R(D,K)$ we can assume that D is a simple divisible R-module. But then by Corollary 8.9 we have $\mathrm{Hom}_R(D,K) \neq 0$.

Let B be a ring extension of R in Q and $B \neq Q$. Then $\mathrm{Hom}_R(B,H) \cong \mathrm{Hom}_R(B,\mathrm{Hom}_R(K,K)) \cong \mathrm{Hom}_R(K \otimes_R B,K) \neq 0$.

Theorem 8.11. *Let $A_1,\ldots,A_n$ be the closed components of R, and let D be a simple divisible submodule of K corresponding to A_1. Then*

(1) $D \subset \bigcap_{j\neq 1} (A_j/R)$

(2) *A_1/R contains every simple divisible submodule of K that is not equal to D.*

Proof. If $j > 1$, then by Corollary 8.9 there is a simple divisible submodule D_j of K corresponding to A_j. By Corollary 1.3 there is an element $f \in \mathrm{Hom}_R(K,K)$ such that $f(K) = D_j$. By Theorem 2.7, $f(D) \subset D$, and thus $f(D) \subset D \cap D_j$. But $D \neq D_j$, and hence $D \cap D_j$ is reduced. Thus $f(D) = 0$ and $D \subset \mathrm{Ker}\ f$. Therefore, $D \subset h(\mathrm{Ker}\ f)$; and since $h(\ker f) = A_j/R$ by Theorem 8.2, we have $D \subset A_j/R$. Thus $D \subset \bigcap_{j\neq 1} (A_j/R)$.

Now if D' is any divisible submodule of K that is not equal to D, we apply the argument of the preceding paragraph with D in the role of D_j and D' in the role of D, and we obtain $D' \subset A_1/R$.

CHAPTER IX

SEMI-SIMPLE AND UNISERIAL DIVISIBLE MODULES

Throughout this chapter R will be a 1-dimensional, local Cohen-Macaulay ring.

Definition. We will say that an Artinian divisible R-module is a semi-simple divisible module if it is a sum of simple divisible R-modules. By analogy with the theory of ordinary semi-simple modules we would expect to find that a semi-simple divisible R-module is isomorphic to a direct sum of simple divisible R-modules. This statement is not correct, but if we replace isomorphism by equivalence it becomes correct, as we shall see in the next theorem.

Theorem 9.1. *Let B be an Artinian divisible R-module. Then the following statements are equivalent:*

(1) *B is a semi-simple divisible R-module.*

(2) *B is equivalent to a direct sum of simple divisible R-modules.*

(3) *If C is any divisible submodule of B, then B is equivalent to $C \oplus B/C$.*

Proof. (1) $\Longrightarrow$ (2). Let B_1 be a simple divisible submodule of B. If $B_1 \neq B$, there is a simple divisible submodule B_2 of B not contained in B_1. Let $D_2 = B_1 + B_2$ and $D_1 = B_1$; then $D_1 \subsetneq D_2$. Continuing in this way we obtain an ascending chain of divisible submodules of B: $0 \subsetneq D_1 \subsetneq D_2 \subsetneq \dots \subsetneq D_i$ such that $D_i = B_1 + B_2 + \dots + B_i$, and every B_j is a simple divisible submodule of B. Since B has ACC on divisible submodules by Theorem 5.5, there is an integer $n > 0$ such that $B = D_n$. Now D_i/D_{i-1} is a nonzero homomorphic image of B_i, and hence is a simple divisible module by Lemma 5.8. Thus we have constructed a composition series of divisible submodules of B, and we see that $L(B) = n$.

We have an exact sequence:

$$0 \to S \to B_1 \oplus \dots \oplus B_n \to B \to 0.$$

Using Theorem 5.11, we have $L(S) = \sum_{i=1}^{n} L(B_i) - L(B) = n - n = 0$, and thus S is a reduced R-module. By Corollary 5.13, B is equivalent to $B_1 \oplus \dots \oplus B_n$.

(2) $\Longrightarrow$ (1). This is a trivial statement.

(1) $\Longrightarrow$ (3). Let C be a divisible submodule of B. Since B has ACC on divisible submodule by Theorem 5.5, there is a divisible submodule A of B that is maximal with respect to the property that $C \cap A$ is reduced. We assert that $A + C = B$. For suppose that this is not the case. Then there is a simple divisible submodule D of B such that $D \not\subset A + C$. Let $A_1 = A + D$; by the maximality of A, $A_1 \cap C$ is not reduced. We have $L((A_1 + C)/A) = L((A + C)/A) + L((A_1 + C)/(A + C)) = L(C/(A \cap C)) + L(D/D \cap (A + C)) = L(C) + 1$. On the other hand we have $L((A_1 + C)/A) = L(A_1/A) + L((A_1 + C)/A_1) = L(D/(A \cap D)) + L(C/(A_1 \cap C)) < L(C) + 1$. This contradiction shows that $A + C = B$.

We have an exact sequence:

$$0 \to A \cap C \to A \oplus C \to B \to 0.$$

Since $A \cap C$ is reduced, we see that B is equivalent to $A \oplus C$ by Lemma 5.8. Now $B/C = (A + C)/C \cong A/(A \cap C)$, and hence A is equivalent to B/C by Lemma 5.8. Combining this with the preceding equivalence we see that B is equivalent to $B/C \oplus C$.

(3) $\Longrightarrow$ (1). Let A be the sum of all of the simple divisible submodules of B. Then by assumption B is equivalent to $A \oplus B/A$. Hence by Corollary 5.13 there is a homomorphism f of $A \oplus B/A$ onto B such that Ker f is reduced. Since f(A) is a sum of simple divisible R-modules, we have $f(A) \subset A$. On the other hand $L(f(A)) = L(A)$ since Ker f is reduced. Thus we have $f(A) = A$.

Suppose that $A \neq B$; then there exists a simple divisible submodule D of B/A and $f(D) \subset A$ and $f(D) \neq 0$. Let $C = f^{-1}(f(D)) \cap A$; then $f(C) = f(D)$ and hence C is not reduced by Theorem 5.4. We have an exact sequence:

$$0 \to S \to C \oplus D \to f(D) \to 0$$

where $S \subset \text{Ker } f$, and hence S is reduced. But then $1 = L(f(D)) = L(C) + L(D) = L(C) + 1$, and hence $L(C) = 0$. Thus C is reduced. This contradiction shows that $B = A$, and hence B is a sum of simple divisible R-modules.

Remarks. We shall prove in a later theorem that every Artinian divisible R-module is a semi-simple divisible R-module if and only if the integral closure of R is a finitely generated R-module. (See the remarks following Corollary 5.14). The same assertion is true for the converse of the following corollary.

Corollary 9.2. *Let* B *be a semi-simple, divisible, Artinian* R-*module, and let* A *be a nonzero, proper divisible submodule of* B. *Then* A *and* B/A *are semi-simple divisible* R-*modules*.

Proof. Clearly every nonzero homomorphic image of B is a semi-simple divisible R-module. By Theorem 9.1 we see that both A and B/A are nonzero homomorphic images of B.

Definition. At the other extreme from semi-simplicity we shall say that an Artinian divisible R-module is *uniserial* if its lattice of divisible submodules is linearly ordered.

Lemma 9.3. *If* D *is a uniserial, divisible, Artinian module and* A *is a submodule of* D, *then* h(A) *and* D/A *are uniserial, divisible, Artinian modules*.

Proof. It is of course trivial that h(A) is a uniserial, divisible Artinian module. Let C/A be a divisible submodule of D/A. Then by Corollary 5.2, $C = h(C) + B$, where B is a finitely generated R-module. Thus $C/A = [(h(C) + A)/A] + [(B + A)/A]$, and $(B + A)/A$ is a finitely generated R-module. Thus by Theorem 5.3, $C/A = (h(C) + A)/A$. From this it follows immediately that D/A is a uniserial, divisible, Artinian module.

<u>Theorem 9.4</u>. *Let D be a uniserial, divisible, Artinian module. Then D is equivalent to a submodule of E and to a factor module of K. Hence $L(D) \leq \min(L(K), L(E))$. Thus there is a finite bound on the lengths of the uniserial, divisible, Artinian modules.*

<u>Proof</u>. Let A be a maximal homomorphic image of K in D. Since D is uniserial, A contains every homomorphic image of K in D. Thus by Corollary 1.3, $A = D$.

Let B be the unique simple divisible submodule of D. Then we have an exact sequence:

$$0 \to B \xrightarrow{i} D \xrightarrow{\pi} D/B \to 0.$$

Hence we have a derived exact sequence:

$$0 \to \mathrm{Hom}_R(D/B, E) \xrightarrow{\pi^*} \mathrm{Hom}_R(D, E) \xrightarrow{i^*} \mathrm{Hom}_R(B, E) \to 0.$$

Since $\mathrm{Hom}_R(B, E) \neq 0$, π^* is not an epimorphism. Thus we can choose $f \in \mathrm{Hom}_R(D, E)$ such that $f \notin \mathrm{Im}\, \pi^*$. To prove the theorem it will be sufficient by Lemma 5.8 to prove that Ker f is reduced. Suppose that this is not the case. Then $B \subset \mathrm{Ker}\, f$ and hence f induces an R-homomorphism $g : D/B \to E$. But then $\pi^*(g) = f$, and we have $f \in \mathrm{Im}\, \pi^*$. This contradiction shows that Ker f is reduced, proving the theorem.

<u>Remark</u>. We shall prove later on that $L(K) = L(E)$. Thus the statement in Theorem 9.4 that $L(D) \leq \min(L(K), L(E))$ is redundant.

<u>Theorem 9.5</u>. *Let B be an R-module equivalent to E. Then B contains one and only one uniserial, divisible, Artinian R-module from*

<u>each equivalence class of such modules</u>.

<u>Proof</u>. <u>Case I: Assume B = E</u>.

By Theorem 9.4, E contains at least one uniserial, divisible, Artinian module from each equivalence class of such modules. Suppose that D_1 and D_2 are equivalent uniserial, divisible, Artinain submodules of E. We will show that $D_1 = D_2$. By definition there is an R-homomorphism f of D_1 onto D_2. Since E is an injective R-module, f can be extended to an R-endomorphism g of E. By Theorem 4.5, $\mathrm{Hom}_R(E,E) = H$; and by Theorem 2.7, D_1, D_2, and E are H-modules. Thus we have $g(D_1) \subset D_1$. However, $g(D_1) = f(D_1) = D_2$. Thus we have $D_2 = g(D_1) \subset D_1$. In similar fashion we have $D_1 \subset D_2$, and thus $D_1 = D_2$.

<u>Case II: General Case: $B \sim E$</u>.

By Corollary 5.13 there is an R-epimorphism f of E onto B such that Ker f is reduced. Thus by Lemma 9.3 and Theorem 9.4, B contains at least one uniserial, divisible, Artinian R-module from each equivalence class of such modules. Suppose that D_1 and D_2 are equivalent uniserial, divisible, Artinian submodules of B. By Corollary 5.13 there is an R-epimorphism g of B onto E such that Ker g is reduced. By Lemma 5.8 and Lemma 9.3, $g(D_1)$ and $g(D_2)$ are equivalent to D_1 and D_2 and are equivalent uniserial, divisible Artinian submodules of E. Thus by Case I we have $g(D_1) = g(D_2)$, and hence $g(D_1 + D_2) = g(D_1)$. Therefore, $L(D_1 + D_2) = L(g(D_1 + D_2))$ $= L(g(D_1)) = L(D_1)$, and thus $D_1 = D_1 + D_2$. Similarly, $D_2 = D_1 + D_2$, and therefore, $D_1 = D_2$.

CHAPTER X

THE INTEGRAL CLOSURE

Throughout this chapter R will be a 1-dimensional, local, Cohen-Macaulay ring.

Theorem 10.1. *Let S be the integral closure of R in Q and let n be the number of distinct pseudo valuation rings of R. Then*

$$L(K) = L(S/R) + n.$$

Proof. By Theorem 5.11 we have $L(K) = L(S/R) + L(Q/S)$. Therefore, it is only necessary to prove that $L(Q/S) = n$. Let $V_1, \ldots, V_n$ be the full set of pseudo valuation rings of R. By Theorem 7.7, $S = \bigcap_{i=1}^{n} V_i$, and by Theorem 7.8, $Q/S \cong Q/V_1 \oplus \ldots \oplus Q/V_n$. By Corollary 8.4 Q/V_i is a simple divisible R-module. Thus $L(Q/S) = n$.

Definition. R is said to be *analytically unramified* if its completion H has no nonzero nilpotent elements. This is equivalent, of course, to asserting that the intersection of the rank 0 prime ideals of H is 0.

Remarks. The equivalent of statements (1) and (2) of the next theorem have been proved for integral domains by Northcott in a different way [19, Th. 8].

Theorem 10.2. *The following statements are equivalent:*

(1) *R is analytically unramified.*

(2) *The integral closure S of R in Q is a finitely generated R-module.*

(3) *$L(K) = n$, where n is the number of distinct pseudo valuation rings of R.*

(4) *K is a semi-simple divisible R-module.*

(5) *E is a semi-simple divisible R-module.*

(6) <u>Every nonzero Artinian divisible R-module is semi-simple.</u>

(7) <u>If</u> C <u>and</u> D <u>are Artinian divisible</u> R-<u>modules, then</u> C <u>and</u> D <u>are equivalent if and only if they have equivalent composition series.</u>

<u>Proof</u>. (1) <==> (2). Let $V_1,\ldots,V_n$ be the distinct pseudo valuation rings of R and $h(V_i/R) = A_i/R$. By Theorem 7.5, $A_1,\ldots,A_n$ are the closed components of R. Let $P_i = \mathrm{Hom}_R(K,A_i/R)$; then $P_1,\ldots,P_n$ are the prime ideals of rank 0 in H by Theorem 7.3. Since $S = \bigcap_i V_i$, we have

$$\bigcap_i P_i = \bigcap_i \mathrm{Hom}_R(K,A_i/R) = \bigcap_i \mathrm{Hom}_R(K,V_i/R) = \mathrm{Hom}_R(K,\bigcap_i (V_i/R))$$
$$= \mathrm{Hom}_R(K,S/R).$$

Thus by Corollary 1.3, $\bigcap_i P_i = 0$ if and only if S/R is reduced. And by Theorem 5.1, S/R is reduced if and only if S is a finitely generated R-module.

(2) <==> (3). This is an immediate consequence of Theorem 10.1 and Theorem 5.1.

(2) ==> (4). If $V_1,\ldots,V_n$ are the distinct pseudo valuation rings of R, then by Theorem 7.8 we have $Q/S \cong Q/V_1 \oplus \ldots \oplus Q/V_n$. Since Q/V_i is a simple divisible R-module by Corollary 8.3, it follows that Q/S is a semi-simple divisible R-module. If S/R is a finitely generated R-module, then K is equivalent to Q/S by Lemma 5.8; and hence K is semi-simple by Theorem 9.1.

(4) ==> (5). By Theorem 5.5, E is a homomorphic image of K^m for some $m > 0$, and thus E is a semi-simple divisible R-module.

(5) ==> (6). A nonzero Artinian divisible R-module is isomorphic to a submodule of E^m for some $m > 0$, by Theorem 5.5; and hence it is a semi-simple divisible R-module by Corollary 9.2.

(6) ==> (7). Let C and D be Artinian divisible R-modules. If they are equivalent, then they have equivalent composition series by Corollary 5.14. Conversely, suppose that they have equivalent compo-

sition series. Then by Theorem 9.1, they will be equivalent to the same finite direct sum of simple divisible modules, and hence, they will be equivalent to each other.

(7) $\Longrightarrow$ (4). Let $D_1,\ldots,D_m$ be the simple divisible factors (with repetitions) of a composition series for K. If $D = D_1 \oplus \ldots \oplus D_m$, then K and D have equivalent composition series, and hence K and D are equivalent by assumption. Therefore K is a semi-simple divisible module by Theorem 9.1.

(4) $\Longrightarrow$ (2). Suppose that S is not a finitely generated R-module. Then by Theorem 5.1, S/R contains a simple divisible R-module D. By Corollary 1.3, there is an epimorphism of K onto D. Let $A/R = h(\operatorname{Ker} f)$; then by Theorem 8.2, A is a closed component of R. By Theorem 8.11, A/R contains every simple divisible submodule of K that is not equal to D. But if V is the pseudo valuation ring corresponding to A, then $D \subset S/R \subset V/R$, and hence $D \subset h(V/R) = A/R$. Thus A/R contains every simple divisible submodule of K. However, this contradicts the fact that K is a semi-simple divisible module, and thus we see that S is a finitely generated R-module.

<u>Theorem 10.3</u>. <u>If</u> R <u>is</u> <u>analytically</u> <u>unramified</u>, <u>then</u> K <u>is</u> <u>equivalent</u> <u>to</u> E.

<u>Proof</u>. Let S be the integral closure of R. By Theorem 10.2 S/R is finitely generated. Hence K is equivalent to Q/S by Lemma 5.8. If $V_1,\ldots,V_n$ are the pseudo valuation rings of R, then by Theorem 7.8 we see that $Q/S \cong Q/V_1 \oplus \ldots \oplus Q/V_n$. Therefore, K is equivalent to $Q/V_1 \oplus \ldots \oplus Q/V_n$.

By Theorem 10.2, E is a semi-simple divisible module, and hence E is equivalent to $D_1 \oplus \ldots \oplus D_t$, where the D_i's are simple divisible R-modules by Theorem 9.1. By Theorem 9.5, D_i is not equivalent to D_j for $i \neq j$, and every equivalence class of simple divisible modules has a representative among the D_i's. Thus $t = n$ and (after renumbering) we have $D_1 \sim Q/V_1$ by Corollary 8.4. Therefore E is equivalent to K.

Remarks. R is said to be a Gorenstein ring if inj. dim. R = 1; that is, if K is an injective R-module. As we shall prove in Theorem 13.1, R is a Gorenstein ring if and only if $K \cong E$. Thus both Gorenstein rings and analytically unramified rings have in common the property that $K \sim E$. This shows that the converse of Theorem 10.3 is false. For there exist Gorenstein rings that are not analytically unramified as we shall see in Theorem 14.16. In Chapter XV we shall find necessary and sufficient conditions for K to be equivalent to E, and we shall see how **closely** these are related to the conditions that R is a Gorenstein ring.

Theorem 10.4. *Assume that R is a Gorenstein ring, or that R is analytically unramified. Then K contains one and only one uniserial, divisible, Artinian R-module from each equivalence class of such modules.*

Proof. Theorem 10.3 and Theorem 9.5 combine to yield an immediate proof.

Remarks. We can now present an extremely simplified proof of a theorem that we proved in [13, Th. 7.4] only with great difficulty.

Theorem 10.5. *Let R be a Noetherian integral domain. Then the following statements are equivalent:*

(1) *R is a Noetherian local domain of Krull dimension 1, and its completion H has only one prime ideal of rank 0.*

(2) *The integral closure of R is a discrete valuation ring.*

(3) *Q/C is an indecomposable R-module for every R-submodule C of Q.*

(4) *R is a Noetherian local domain of Krull dimension 1 and Q/I is an indecomposable R-module for every ideal I of R.*

Proof. (1) $\Longleftrightarrow$ (2). If the integral closure of R is a discrete valuation ring, then R is a 1-dimensional Noetherian local domain. By Theorem 7.4, there is a one-to-one correspondence between the valuation rings of R and the prime ideals of rank 0 in H. This establishes the equivalence of (1) and (2) immediately.

(2) $\Longrightarrow$ (3). Let V be the integral closure of R, and $h(V/R) = A/R$. Since V is a valuation ring, A is a closed component of R by Theorem 7.5; and it is the only closed component of R. Since every proper divisible submodule D of K is contained in a maximal proper divisible submodule of K, and since A/R is the only maximal proper divisible submodule of K, we have $D \subset A/R$.

Let C be a proper, nonzero R-submodule of Q, and suppose that $Q/C = B_1/C \oplus B_2/C$, where B_1, B_2 are R-submodules of Q. Then $B_1 + B_2 = Q$ and $B_1 \cap B_2 = C$. Without loss of generality, we can assume that $R \subset C$. Thus we have $B_1/R + B_2/R = K$. Let $D_1 = h(B_1/R)$ and $D_2 = h(B_2/R)$; then by Corollary 5.2, $B_1/R = D_1 + T_1$ and $B_2/R = D_2 + T_2$, where T_1 and T_2 are finitely generated submodules of K. Hence we have $K = D_1 + D_2$ by Theorem 5.3. If both D_1 and D_2 are proper submodules of K, then we have $K = D_1 + D_2 \subset A/R$ by the preceding paragraph. This contradiction shows that we can assume that $D_1 = K$. But then $B_1 = Q$, and this shows that Q/C is an indecomposable R-module.

(3) $\Longrightarrow$ (4). Of course it is trivial that Q/I is indecomposable for every ideal I of R. Suppose that R is not a 1-dimensional local ring. Then R has two distinct rank 1 prime ideals I_1 and I_2, and we put $\mathcal{S} = R - (I_1 \cup I_2)$ and $C = R_{\mathcal{S}}$. Then C is a Noetherian 1-dimensional domain with two distinct maximal ideals I_1C and I_2C. By Theorem 4.1 we have: $Q/C \cong Q/C_{I_1} \oplus Q/C_{I_2}$, a nontrivial direct sum decomposition. This contradiction shows that R is a Noetherian, local, 1-dimensional domain.

(4) $\Longrightarrow$ (2). Let S be a ring extension of R in Q that is

finitely generated as an R-module. If S is not a local ring, then Q/S has a non-trivial direct sum decompostion by Theorem 4.1. Since S is isomorphic to an ideal of R, this contradiction shows that S is a local ring.

Let V be the integral closure of R, and suppose that V is not a discrete valuation ring. Then there exist non-units x and y in V such that $x + y = 1$. Let $S = R[x,y]$; then S is a finitely generated R-module, since x and y are integral over R. Thus S is a local ring, and hence either x or y is a unit in S. But then either x or y is a unit in V, and this contradiction shows that V is a discrete valuation ring.

CHAPTER XI

THE PRIMARY DECOMPOSITION

Throughout this chapter R will be a 1-dimensional, local, Cohen-Macaulay ring.

Definition. Let D be an Artinian, divisible R-module and P a prime ideal of rank 0 in H. We shall say that D is a P-primary module if $\mathrm{Ann}_H D$ is a P-primary ideal, or if D = 0.

Theorem 11.1. Let D be an Artinian, divisible R-module and P a prime ideal of rank 0 in H. Then D is a P-primary module if and only if all of the simple divisible factor modules in a composition series for D correspond to P.

Proof. Suppose that D is a P-primary module and let $0 = D_0 \subset D_1 \subset \dots \subset D_n = D$ be a composition series of divisible submodules of D. Then $\mathrm{Ann}_H D \subset \mathrm{Ann}_H(D_{i+1}/D_i)$ and $\mathrm{Ann}_H(D_{i+1}/D_i)$ is a prime ideal of rank 0 in H by Corollary 8.3.

Since $\mathrm{Ann}_H D$ is a P-primary ideal, it follows that $\mathrm{Ann}_H(D_{i+1}/D_i) = P$ for all $i = 0,\dots,n-1$; and hence D_{i+1}/D_i corresponds to P.

Conversely, suppose that every simple divisible factor module of a composition series of divisible submodules of D corresponds to P. Let $I = \mathrm{Ann}_H D$; and we shall prove that I is a P-primary ideal. Since $PD_{i+1} \subset D_i$ for all i, we have $P^n D = 0$, and thus $P^n \subset I$. Because $ID_{i+1} \subset D_i$, we have $I \subset P$. Let f and g be elements of H such that $fg \in I$ and $f \notin P$. We assert that $fD = D$. For if $fD \neq D$, then there is a composition series of divisible submodules of D passing through fD by Theorem 5.10. Hence there exists a divisible submodule B of D such that $fD \subset B$ and $C = B/fD$ is a simple divisible R-module corresponding to P. But $fC = 0$, and hence $f \in P$. This contradiction shows that $fD = D$. We then have $0 = (gf)D = gD$, and hence $g \in I$. This proves that I is a P-primary ideal.

Corollary 11.2. Let D be a nonzero P-primary divisible Artinian R-module, where P is a prime ideal of rank 0 in H, and let $f \in H$. Then $fD = D$ if and only if $f \notin P$.

Proof. If $f \in P$, then there exists an integer $n > 0$ such that $f^n \in \mathrm{Ann}_H D$, and hence $fD \neq D$. Conversely, if $f \notin P$, then we have shown that $fD = D$ in the course of the proof of Theorem 11.1.

Corollary 11.3. Let $0 \to B \to C \to D \to 0$ be an exact sequence of Artinian, divisible R-module, and let P be a prime ideal of rank 0 in H. Then C is a P-primary module if and only if B and D are P-primary modules.

Proof. By Theorem 5.10, there is a composition series of divisible submodules of C passing through B. Hence the set of simple divisible factor modules for C is the union of the sets of simple divisible factor modules for B and for D. The corollary now follows directly from Theorem 11.1.

Corollary 11.4. (1) If $D_1, \ldots, D_n$ are Artinian, divisible R-modules and P is a prime ideal of rank 0 in H, then $D_1 \oplus \ldots \oplus D_n$ is a P-primary module if and only if every D_i is a P-primary module.

(2) If D is an Artinian, divisible R-module, and B and C are P-primary divisible submodules of D, then $B + C$ is also a P-primary divisible submodule of D.

(3) If D is an Artinian, divisible R-module, then D contains a unique largest P-primary divisible submodule C; C contains every divisible P-primary submodule of D; and D/C has no nonzero P-primary divisible submodules.

(4) If B and B' are P-primary and P'-primary Artinian divisible R-modules respectively, with $P \neq P'$, then $\mathrm{Hom}_R(B,B') = 0$.

Proof. (1) follows immediately from Corollary 11.3 and (2) follows from (1) and Corollary 11.3. (3) now follows from (2) and Corollary 11.3. As for (4), let $f \in \mathrm{Hom}_R(B,B')$, and suppose that f is not zero. Then $f(B)$ is both P-primary and P'-primary by Corollary 11.3. This contradiction shows that $\mathrm{Hom}_R(B,B') = 0$.

Definition. Let D be an Artinian, divisible R-module and P a prime ideal of rank 0 in H. We shall call the unique largest P-primary divisible submodule of D that is guaranteed to exist by Corollary 11.4 the P-primary component of D. We shall denote this component by D_P.

Theorem 11.5. Let D be an Artinian, divisible R-module, and let $P_1,\ldots,P_n$ be the prime ideals of rank 0 in H. Then:

(1) $D = D_{P_1} + \ldots + D_{P_n}$.

(2) D is equivalent to $D_{P_1} \oplus \ldots \oplus D_{P_n}$.

Proof. (1) Let $C = D_{P_1} + \ldots + D_{P_n}$ and suppose that $C \neq D$. Then there is a divisible submodule B of D such that $C \subset B$ and B/C is a simple divisible R-module. We can assume that B/C corresponds to P_1. Let $A = D_{P_2} + \ldots + D_{P_n}$; then C/A is a homomorphic image of D_{P_1}; and thus by Corollary 11.3, C/A is a P_1-primary module. We have an exact sequence:

$$0 \to C/A \to B/A \to B/C \to 0$$

where the extremes are P_1-primary modules. Hence by Corollary 11.3, B/A is a P_1-primary module.

Let $I = \mathrm{Ann}_H A$; then $\bigcap_{i \geq 2} \mathrm{Ann}_H D_{P_i} \subset I$. Since $\mathrm{Ann}_H D_{P_i}$ is a P_i-primary ideal, (or is equal to H in the case that $D_{P_i} = 0$) we see that $I \not\subset P_1$. Choose $h \in I$ such that $h \notin P_1$. Because B is an H-module by Theorem 2.7, h induces an R-homomorphism $f : B/A \to B$ defined by $f(x + A) = hx$ for all $x \in B$. We assert that Ker f is a reduced R-module.

Suppose that Ker f is not reduced; then Ker f contains a simple divisible R-module F/A. Since B/A is P_1-primary we have $\mathrm{Ann}_H F/A = P_1$. If $x \in F$, then $hx = f(x + A) = 0$, and thus $h \in \mathrm{Ann}_H F \subset \mathrm{Ann}_H F/A = P_1$,

But $h \notin P_1$, and this contradiction shows that Ker f is reduced.

Let $G = f(B/A) \subset B$; then G is a P_1-primary module by Corollary 11.3, and hence $G \subset D_{P_1} \subset C$. By Corollary 11.2 we have $G = hG$, and thus $f(B/A) = f((G + A)/A)$. Therefore, we have $B/A = (G + A)/A + \text{Ker } f$. Since Ker f is reduced, we have $B = G + A$ by Theorem 5.3. Thus $B \subset C$, and this is a contradiction. Hence we have $D = D_{P_1} + \dots + D_{P_n}$.

(2) Now $\text{Ann}_H(D_{P_2} + \dots + D_{P_n}) \supset \bigcap_{i \geq 2} \text{Ann}_H D_{P_i}$, and hence $D_{P_2} + \dots + D_{P_n}$ has no nonzero P_1-primary component. Thus $D_{P_1} \cap (D_{P_2} + \dots + D_{P_n})$ is a reduced R-module. By statement (1) we have an exact sequence:

$$0 \to D_{P_1} \cap (D_{P_2} + \dots + D_{P_n}) \to D_{P_1} \oplus (D_{P_2} + \dots + D_{P_n}) \to D \to 0.$$

Hence by Lemma 5.8, D is equivalent to $D_{P_1} \oplus (D_{P_2} + \dots + D_{P_n})$. By induction on the number of nonzero primary components, $D_{P_2} + \dots + D_{P_n}$ is equivalent to $D_{P_2} \oplus \dots \oplus D_{P_n}$. Thus D is equivalent to $D_{P_1} \oplus (D_{P_2} \oplus \dots \oplus D_{P_n})$.

Corollary 11.6. *A uniserial, Artinian, divisible R-module is P-primary for some prime ideal P of rank 0 in H.*

Proof. This is an immediate consequence of Theorem 11.5.

Remarks. It is clear from Theorem 11.5 that if D is an Artinian divisible R-module, then a composition series of divisible submodules of D can be found so that all of the simple divisible factor modules corresponding to a given prime ideal P of rank 0 in H can be arranged in sequence in a single block, and that the position of these blocks in the chain is arbitrary.

Corollary 11.7. *If A is an Artinian divisible R-module, then a given simple divisible R-module D is isomorphic to a factor module in a composition series of divisible submodules of A if and only if*

D is equivalent to a submodule of A.

Proof. This is clear from the preceding remarks about Theorem 11.5.

Theorem 11.8. Let $0 = N_1 \cap \dots \cap N_n$ *be a normal decomposition of* 0 *in* H, *where* N_i *is a* P_i*-primary ideal of* H *and* $P_1, \dots, P_n$ *are the prime ideals of rank* 0 *in* H. *Let* $K \otimes_R N_i = B_i/R \subset K$, *and let* $U = B_1 \cap \dots \cap B_n$. *Then:*

(1) B_i *is a strongly unramified ring extension of* R *in* Q.

(2) $B_i + \bigcap_{j \neq i} B_j = Q$, *and thus* $Q/U \cong Q/B_1 \oplus \dots \oplus Q/B_n$.

(3) *If* A_i *is the closed component of* R *corresponding to* P_i, *then* $B_i \subset A_i$ *and* $B_i \not\subset A_j$ *for* $j \neq i$.

(4) U *is a finitely generated* R*-module, and thus* U *is a 1-dimensional, semi-local, Cohen Macaulay ring. If* R *is an integral domain, then the ideals* $M_i = B_iM \cap U$ *are all distinct and are the only maximal ideals of* U, *and* $B_i = U_{M_i}$.

(5) K *is equivalent to* $Q/B_1 \oplus \dots \oplus Q/B_n$.

(6) Q/B_i *is isomorphic to the* P_i*-primary component of* Q/U *and* Q/B_i *is equivalent to the* P_i*-primary component of* K.

(7) $K \otimes_R (\bigcap_{j \neq i} N_j) = h(\bigcap_{j \neq i} B_j/R)$ *is the* P_i*-primary component of* K.

Proof. (1) B_i is a strongly unramified extension of R in Q by Theorem 6.6.

(2) Let $j_1, \dots, j_k$ be any set of integers between 1 and n; and let A be a closed component of R such that A/R contains $h(\bigcap_{j=1}^{k} B_{j_t}/R)$. By Corollary 6.7, $h(\bigcap_{j=1}^{k} B_{j_t}/R) = K \otimes_R (\bigcap_{j=1}^{k} N_{j_t})$; and by Theorem 7.3, $A/R = K \otimes_R P$, where P is the prime ideal of rank 0 in H corresponding to A. Then by Theorem 6.6, $\bigcap_{t=1}^{k} N_{j_t} \subset P$. Hence $P = P_{j_t}$ for some integer j_t and thus $A = A_{j_t}$. Therefore, if j is an integer, $1 \leq j \leq n$, such that j is not equal to any j_t then there is no closed

component A of R such that A/R contains $(B_j/R) + h(\bigcap_{j=1}^{k} B_{j_t}/R)$. Hence we have $B_j + \bigcap_{t=1}^{k} B_{j_t} = Q$. It follows from this that $Q/U \cong Q/B_j + Q/(\bigcap_{i\neq j} B_i)$. By induction on n we have $Q/(\bigcap_{i\neq j} B_i) \cong \sum_{i\neq j} \oplus Q/B_i$, and thus $Q/U \cong \sum_{i=1}^{n} \oplus Q/B_i$.

(3) We have proved in (2) that if A_i is the closed component of R corresponding to P_i, then A_i is the only closed component of R containing B_i.

(4) By Corollary 6.7 we have $0 = K \otimes_R (\bigcap N_i)$ $= h(\bigcap_i (K \otimes_R N_i)) = h(\bigcap_i B_i/R) = h(U/R)$, and thus U/R is a reduced R-module. Therefore, by Theorem 5.1, U is a finitely generated R-module; and hence U is a semi-local, 1-dimensional, Cohen-Macaulay ring. If R is an integral domain, then Theorem 6.3 shows that the ideals $M_i = B_iM \cap R$ are exactly the set of maximal ideals of U and that $B_i = U_{M_i}$.

(5) Since U/R has finite length by (4), it follows from the exact sequence:

$$0 \to U/R \to K \to Q/U \to 0$$

and Lemma 5.8 that K is equivalent to Q/U.

(6) By Theorem 6.6, $\mathrm{Ann}_H(Q/B_i) = N_i$, and thus Q/B_i is the P_i-primary component of $Q/B_1 \oplus \ldots \oplus Q/B_n$. Hence Q/B_i is isomorphic to the P_i-primary component of Q/U and equivalent to the P_i-primary component of K.

(7) By (2) we have $B_i + \bigcap_{j\neq i} B_j = Q$, and thus $Q/B_i \cong \bigcap_{j\neq i} B_j/U$. Therefore, by (6), $\bigcap_{j\neq i} B_j/U$ is the P_i-primary component of Q/U. We have an exact sequence:

$$0 \to U/R \to \bigcap_{j\neq i} B_j/R \to \bigcap_{j\neq i} B_j/U \to 0.$$

Hence by Theorem 5.3 and Lemma 5.8 we see that $h(\bigcap_{j\neq i} B_j/R)$ is equivalent to $\bigcap_{j\neq i} B_j/U$, and thus to Q/B_i. Since Q/B_i is equivalent to the P_i-primary component of K by (6), it follows that $h(\bigcap_{j\neq i} B_j/R)$ is the P_i-primary component of K. By Corollary 6.7, we have
$h(\bigcap_{j\neq i} B_j/R) = K \otimes_R (\bigcap_{j\neq i} N_j)$.

Definition. With the notation of Theorem 11.8, we shall say that B_i is the P_i-*primary component* of R.

Corollary 11.9. *With the notation of Theorem* 11.8 *we have an exact sequence*

$$0 \to H \to H/N_1 \oplus \dots \oplus H/N_n \to U/R \to 0$$

where U/R *has finite length and* $H_1/N_1 \oplus \dots \oplus H/N_n \cong H(U)$, *the completion of* U *in the Zariski topology*.

Proof. By Theorem 6.4 and Theorem 11.8, it is sufficient to prove that $H/N_1 \oplus \dots \oplus H/N_n \cong H(U)$. Now $Q/U \cong Q/B_1 \oplus \dots \oplus Q/B_n$ by Theorem 11.8 and $\mathrm{Hom}_R(Q/B_i, Q/B_j) = 0$ for $i \neq j$ by Corollary 11.4. Thus $H(U) \cong \mathrm{Hom}_R(Q/S,Q/S) \cong \sum_i \oplus \mathrm{Hom}_R(Q/B_i, Q/B_i)$. By Theorem 6.6 we have $\mathrm{Hom}_R(Q/B_i, Q/B_i) \cong H/N_i$, and thus $H(U) \cong \sum_i \oplus H/N_i$.

Theorem 11.10. *Let* A *and* B *be Artinian divisible* R-*modules. Then* A *and* B *have equivalent composition series if and only if* $L(A_p) = L(B_p)$ *for all prime ideals* P *of rank* 0 *in* H. *In this case we have* $L(A) = L(B)$.

Proof. This is an immediate consequence of Theorem 11.5.

CHAPTER XII

THE FIRST NEIGHBORHOOD RING

Throughout this chapter R will be a 1-dimensional, local Cohen-Macaulay ring. If A is an R-module of finite length, we shall denote its length by $\mathcal{L}(A)$. For all large values of n, $\mathcal{L}(R/M^n)$ is a polynomial in n of degree 1 called the Hilbert polynomial of R [24, VIII, Th. 20]. Thus $\mathcal{L}(R/M^n) = en - \rho$ for large n, where e and ρ are integers called the multiplicity and reduction number of R, respectively ([24, VIII, §10] and [22]). It follows that $\mathcal{L}(M^n/M^{n+1}) = e$ is constant for large n, and from this we see that e is a positive integer. We also note that ρ is a nonnegative integer as we shall see later. The purpose of this chapter is to study some of the relationships between e and ρ.

An element $a \in M^s$ is called superficial of degree s if $(M^{n+s}: a) = M^n$ for all large n. For every large value of s there is a superficial element of degree s [20, Th. 2]. Northcott has proved that if $a \in M^s$, then a is superficial of degree s if and only if $M^n a = M^{n+s}$, for all large n [20, Th. 1]. We shall consistently use this equation as the defining property of superficial elements. Clearly a superficial element is a regular element of R, and hence generates an M-primary ideal.

Northcott has defined the first neighborhood ring Λ of R to be the set of all elements (in the full ring of quotients Q of R) of the form b/a, where $b \in M^s$ and a is superficial of degree s (s variable) [20]. Since the set of superficial elements of R is obviously multiplicatively closed, it is clear that Λ is indeed a subring of Q containing R. We shall use the symbol Λ for the first neighborhood ring without further definition.

Northcott's pioneering work forms the basis of any treatment of the subject. We shall have frequent occasion to make use of certain of his results and we shall collect them here in a single theorem.

We shall prove these results in order to make this chapter as self-contained as possible, and also because it has been possible to find somewhat simpler proofs than the original ones (see [20] and [22]).

Theorem 12.1 (Northcott's Theorem). (1) *There exists a superficial element* a *of degree* s *such that* $\Lambda = M^s a^{-1}$. *Thus* Λ *is a finitely generated* R-*module, and hence is a semi-local*, 1-*dimensional Cohen-Macaulay ring*.

(2) $\Lambda M^n = M^n$ *for all large* n.

(3) *If* $b \in M^t$, *then* b *is superficial of degree* t *if and only if* $\Lambda b = \Lambda M^t$.

(4) $\Lambda = R$ *if and only if* R *is a discrete valuation ring*.

Proof. (1) Suppose that b is a superficial element of degree t in R. Choose an integer k large enough so that $M^{kt}b = M^{kt+t}$. If we let $a = b^k$ and $s = kt$, then $M^s a = M^{2s}$, and a is superficial of degree s. Putting $T = M^s a^{-1}$, we see that $T^2 = T$, and thus T is a ring contained in Λ. Since T is a finitely generated R-module, and $M^t b^{-1} \subset T$, we see that every element of $M^t b^{-1}$ is integral over R. However, b was an arbitrary superficial element of R, and thus we have proved that Λ is an integral extension of R.

On the other hand, let $y \in \Lambda$; then $y = d/c$, where $d \in M^m$ and c is superficial of degree m. We write $y = (dc^{s-1}/a^m)(a^m/c^s)$ and note that dc^{s-1}/a^m and c^s/a^m are elements of T. Since c^s/a^m is clearly a unit in Λ, and Λ is integral over T, it is also a unit in T. Thus we see that $y \in T$, and we have proved that $\Lambda = T$.

(2) By (1) we have $\Lambda a = M^s$, and hence $\Lambda M^s = M^s$. Therefore, $\Lambda M^n = M^n$ for all $n \geq s$.

(3) If b is superficial of degree t, then by (1), $\Lambda M^t = (M^{kt}/b^k)M^t = M^{kt+t}/b^k = (M^{kt}b)/b^k = \Lambda b$. Conversely, suppose that $b \in M^t$ and that $\Lambda b = \Lambda M^t$. Then by (2) we have for all large n that $M^n b = \Lambda M^n b = \Lambda M^{n+t} = M^{n+t}$, and thus b is superficial of degree t.

(4) If R is a discrete valuation ring, then R is integrally closed, and hence by (1) we have $\Lambda = R$. Conversely, suppose that $\Lambda = R$. Then by (1) we have $M^s = \Lambda a = Ra$. Thus there exist elements $u_1,\dots,u_m$ in $M^{s-1}a^{-1}$ and elements $b_1,\dots,b_m$ in M such that $1 = \sum_{i=1}^{m} u_i b_i$ and $u_i M \subset R$ for all i. Since R is a local ring, there is an index j such that $1 = u_j b_j$, and hence $M = Rb_j$ is a principal ideal of R. It follows readily that R is a discrete valuation ring.

Our first proposition is an elementary generalization of a result of Northcott [21, Th. 1], but nevertheless, it is of crucial importance.

Theorem 12.2. ΛM is a principal ideal of Λ, and is generated by a regular element of Λ.

Proof. Let $N_1,\dots,N_t$ be the set of maximal ideals of Λ, and let $J = N_1 \cap \dots \cap N_t$. Then J is the Jacobson radical of Λ. If we denote ΛM by I, then by the Chinese Remainder Theorem we have:

$$I/JI \cong I/N_1 I \oplus \dots \oplus I/N_t I.$$

We shall prove that $I/JI \cong \Lambda/J$. Since $\Lambda/J \cong \Lambda/N_1 \oplus \dots \oplus \Lambda/N_t$ by the Chinese Remainder Theorem, it is sufficient to prove that $I/N_i I \cong \Lambda/N_i$ for all $i = 1,\dots,t$.

By Theorem 12.1, there is an element $a \in M^s$ such that $\Lambda a = M^s$. Thus $\Lambda_{N_i} a = \Lambda_{N_i} M^s = (I_{N_i})^s$, and hence there exist elements $u_1,\dots,u_m$ in $(I_{N_i})^{s-1}a^{-1}$ and $y_1,\dots,y_m$ in I_{N_i} such that $1 = \sum_{k=1}^{m} u_k y_k$ and $u_k I_{N_i} \subset \Lambda_{N_i}$. Since Λ_{N_i} is a local ring there is an index j such that $1 = u_j y_j$. Therefore, $I_{N_i} = \Lambda_{N_i} y_j$; and since I_{N_i} contains a regular element of Λ_{N_i} we have $I_{N_i} \cong \Lambda_{N_i}$ for every i. Thus we have $I/N_i I \cong (I/N_i I)_{N_i} \cong I_{N_i}/N_i I_{N_i} \cong \Lambda_{N_i}/N_i \Lambda_{N_i} \cong \Lambda/N_i$ for all i, and this proves that $I/JI \cong \Lambda/J$.

Because I/JI is a cyclic Λ-module, there is an element x in I such that $I = \Lambda x + JI$. However, J is the Jacobson radical of Λ and hence by the Nakayama Lemma we have $I = \Lambda x$. Since I contains a regular element of R, it follows immediately that x is a regular element of Λ.

Theorem 12.3. Suppose that $a \in M$ is a regular element of R and that I and J are finitely generated R-submodules of the full ring of quotients of R such that both I and J contain regular elements of R. Then I/aI and J/aJ have finite length and $\mathcal{L}(I/aI) = \mathcal{L}(J/aJ)$.

Proof. We may assume without loss of generality that I is an ideal of R. Since $Ra \cap I$ contains a regular element of R, it is an M-primary ideal, and hence there is an integer $n > 0$ such that $M^n \subset Ra \cap I$. Thus R/I and I/aI have finite length. From the exact sequence:

$$0 \to I/aI \to R/aI \to R/I \to 0$$

we conclude that $\mathcal{L}(R/aI) = \mathcal{L}(I/aI) + \mathcal{L}(R/I)$. However, we have another exact sequence:

$$0 \to Ra/Ia \to R/aI \to R/Ra \to 0$$

from which we conclude that $\mathcal{L}(R/aI) = \mathcal{L}(R/Ra) + \mathcal{L}(Ra/Ia)$. But we have $Ra/Ia \cong R/I$ since a is a regular element of R, and thus comparing the two equations for $\mathcal{L}(R/aI)$ we see that $\mathcal{L}(I/aI) = \mathcal{L}(R/Ra)$. In similar fashion we have $\mathcal{L}(J/aJ) = \mathcal{L}(R/Ra)$; and thus $\mathcal{L}(I/aI) = \mathcal{L}(J/aJ)$.

Northcott has proved that $\mathcal{L}(\Lambda/R) = \rho$, thus showing that $\rho \geq 0$ [28, Th. 1]. In the next proposition we shall give a proof of this result, and in addition simultaneously prove that $\mathcal{L}(\Lambda/\Lambda M) = e$.

Theorem 12.4. $\mathcal{L}(\Lambda/\Lambda M) = e$ and $\mathcal{L}(\Lambda/R) = \rho$. Furthermore, $\mathcal{L}(\Lambda/\Lambda M^n) = en$ for all $n > 0$.

Proof. Let $\mathcal{L}(\Lambda/\Lambda M) = \alpha$ and $\mathcal{L}(\Lambda/R) = \beta$. By Theorem 12.2 there is a regular element x of Λ such that $\Lambda x = \Lambda M$. Thus $\Lambda M^i/\Lambda M^{i+1} \cong \Lambda/\Lambda x$ for every i, and hence $\mathcal{L}(\Lambda/\Lambda M^n) = n\alpha$ for every $n \geq 0$. For all large k we have by Theorem 12.1 that $\Lambda M^k = M^k$ and $\mathcal{L}(R/M^k) = ek - \rho$. Thus $\alpha k = \mathcal{L}(\Lambda/\Lambda M^k) = \mathcal{L}(\Lambda/M^k) = \mathcal{L}(\Lambda/R) + \mathcal{L}(R/M^k) = \beta + (ek - \rho)$. Letting $k \to \infty$, we see that $\alpha = e$ and $\beta = \rho$.

<u>Remarks</u>. Let $N_1, \ldots, N_t$ be the set of maximal ideals of Λ and let $\Lambda_i = \Lambda_{N_i}$. Then $\Lambda \cap \Lambda_i M$ is an N_i-primary ideal of Λ and $\Lambda M = \bigcap_{i=1}^{t} (\Lambda \cap \Lambda_i M)$ is a normal decomposition of ΛM. Since $\Lambda/(\Lambda \cap \Lambda_i M) \cong \Lambda_i/\Lambda_i M$, we have $\Lambda/\Lambda M \cong \sum_{i=1}^{t} \oplus \Lambda_i/\Lambda_i M$ by the Chinese Remainder Theorem. Hence we see that $\mathcal{L}(\Lambda/\Lambda M) = \sum_{i=1}^{t} \mathcal{L}(\Lambda_i/\Lambda_i M)$. If we let η_i be the length of $\Lambda_i/\Lambda_i M$ as a Λ_i-module and $\lambda_i = [\Lambda_i/\Lambda_i N_i : R/M]$, and note that by Theorem 12.4 $\mathcal{L}(\Lambda/\Lambda M) = e$, then the preceding equation becomes $e = \sum_{i=1}^{t} \lambda_i \eta_i$, a formula due to Northcott [21, Th. 2].

The next proposition is a strengthened version of [21, Proposition 1].

<u>Theorem 12.5</u>. *Let $a \in M^s$ and let I be a finitely generated R-submodule of Q such that I contains a regular element of R. Then $\mathcal{L}(I/aI) \geq se$ and $\mathcal{L}(I/aI) = se$ if and only if a is superficial of degree s.*

<u>Proof</u>. If a is a zero divisor in R, then $\mathcal{L}(I/aI)$ is infinite, and thus without loss of generality we may assume that a is a regular element of R. Hence by Theorem 12.3 we may assume that $I = \Lambda$. Since $\Lambda a \subset \Lambda M^s$, we have $\mathcal{L}(\Lambda/\Lambda a) \geq se$ and $\mathcal{L}(\Lambda/\Lambda a) = se$ if and only if $\Lambda a = \Lambda M^s$ by Theorem 12.4. By Theorem 12.1, $\Lambda a = \Lambda M^s$ if and only if a is superficial of degree s.

<u>Corollary 12.6.</u> *If $a \in M^s$ and $b \in M^t$, then ab is superficial of degree s + t if and only if a is superficial of degree a and b is*

<u>superficial</u> <u>of</u> <u>degree</u> t.

<u>Proof</u>. If a is superficial of degree s and b is superficial of degree t, then for large n, $M^n ab = M^{n+s}b = M^{n+s+t}$, and hence ab is superficial of degree s + t. Conversely, suppose that ab is superficial of degree s + t. Then $\mathcal{L}(\Lambda/\Lambda ab) = (s + t)e$ by Theorem 12.5. Now b is a regular element of R and thus $\Lambda a/\Lambda ab \cong \Lambda/\Lambda b$. Hence $se + te = \mathcal{L}(\Lambda/\Lambda ab) = \mathcal{L}(\Lambda/\Lambda a) + \mathcal{L}(\Lambda a/\Lambda ab) = \mathcal{L}(\Lambda/\Lambda a) + \mathcal{L}(\Lambda/\Lambda b)$. It follows from Theorem 12.5 that a is superficial of degree s and b is superficial of degree t.

<u>Definition</u>. Let X be an indeterminate over R; then by [18, Ch. IV., Prop. 5], M[X] is a prime ideal of rank 1 in R[X] and we can define $R^* = R[X]_{M[X]}$. We have a natural imbedding $R \subset R^*$; and if I is an ideal of R, we define $I^* = R^*I$. Then we have $MI^* = (MI)^* = M^*I^*$. R* is a local, 1-dimensional Cohen-Macaulay ring. If I is an M-primary ideal of R, then by [18, Ch. IV, Cor. 1 and Prop. 2], I* is an M*-primary ideal of R* and $I^* \cap R = I$. If A* is an R*-module of finite length, we shall denote its length by $\mathcal{L}^*(A^*)$. We shall denote the multiplicity and reduction number of R* by e* and ρ^*, respectively.

The following theorem was communicated to me by D. Rees.

<u>Theorem 12.7</u> (D. Rees). (1) <u>If</u> $I \subset J$ <u>are</u> <u>M-primary</u> <u>ideals</u> <u>of</u> R, <u>then</u> $\mathcal{L}^*(J^*/I^*) = \mathcal{L}(J/I)$.

(2) $e^* = e$ <u>and</u> $\rho^* = \rho$.

(3) R*/M* <u>is</u> <u>infinite,</u> <u>and</u> <u>thus</u> R* <u>has</u> <u>a</u> <u>superficial</u> <u>element</u> <u>of</u> <u>degree</u> 1.

<u>Proof</u>. If $\mathcal{L}(J/I) = n$, there exists a chain of M-primary ideals $I = I_0 \subset I_1 \subset \ldots \subset I_n = J$ such that $I_k/I_{k-1} \cong R/M$ for all $k = 1,\ldots,n$. Thus for each k there is an element $a_k \in I_k$ such that $I_k = I_{k-1} + Ra_k$ and $Ma_k \subset I_{k-1}$. Therefore, $I^* = I_0^* \subset I_1^* \subset \ldots \subset I_n^* = J^*$ is a chain of M*-primary ideals of R*; and since $I_k^* \cap R = I_k$,

the I_k^* are all distinct. We have $I_k^* = I_{k-1}^* + R^*a_k$ and $M^*a_k = (Ma_k)^* \subset I_{k-1}^*$. Thus $I_k^*/I_{k-1}^* \cong R^*/M^*$, and the I_k^* gives us a composition series from I^* to J^*. Therefore, $\mathcal{L}^*(J^*/I^*) = \mathcal{L}(J/I)$.

If $j > 0$ is any integer, then $\mathcal{L}^*((M^*)^j/(M^*)^{j+1})$ $= \mathcal{L}^*((M^j)^*/(M^{j+1})^*) = \mathcal{L}(M^j/M^{j+1})$ by the preceding paragraph. From this it follows immediately that $e^* = e$ and $\rho^* = \rho$.

It is obvious that R^*/M^* is infinite, and thus by [24, p. 287], R^* has a superficial element of degree 1.

I am indebted to D. Rees for suggesting the method of proof and the strengthening of the statement of the next theorem. It is a special case of a more general (unpublished) result of his concerning ideals of height 1 in a Cohen-Macaulay ring.

Theorem 12.8. *Every M-primary ideal of R can be generated by e elements.*

Proof. If I is an M-primary ideal of R, then the minimal number of generators of I is equal to $\mathcal{L}(I/MI)$, and we wish to prove that $\mathcal{L}(I/MI) \leq e$. Because of Theorem 12.7 we can assume without loss of generality that R has a superficial element a of degree 1. But then $aI \subset MI$, and hence $\mathcal{L}(I/MI) \leq \mathcal{L}(I/aI) = e$ by Theorem 12.5.

We have noted in the proof of Theorem 12.7 that if R/M is infinite, then R has a superficial element of degree 1. Another instance in which the existence of such an element is guaranteed is given by the following theorem.

Theorem 12.9. *If Λ is a local ring, then R has a superficial element of degree 1.*

Proof. By Theorem 12.2 there is an element x in Λ such that $\Lambda M = \Lambda x$. Now M is generated by elements $a_1, \ldots, a_n$ over R, and hence there exist elements $u_1, \ldots, u_n$ in Λ such that $x = \sum_{i=1}^{n} u_i a_i$. On the other hand there exist elements $v_1, \ldots, v_n$ in Λ such that $a_i = v_i x$.

Thus $x = (\sum_{i=1}^{n} u_i v_i)x$, and hence $1 = \sum_{i=1}^{n} u_i v_i$. Because Λ is a local ring there is an index j, $1 \leq j \leq n$ such that v_j is a unit in Λ. Therefore, $\Lambda M = \Lambda x = \Lambda v_j x = \Lambda a_j$, and thus a_j is a superficial element of R of degree 1 by Theorem 12.1.

<u>Theorem 12.10</u>. <u>There exists an integer</u> $\nu \geq 0$ <u>such that</u> $\mathcal{L}(M^n/M^{n+1}) = e$ <u>for all</u> $n \geq \nu$ <u>and</u> $\mathcal{L}(M^n/M^{n+1}) < e$ <u>if</u> $n < \nu$.

<u>Proof</u>. Because of Theorem 12.7 we can assume without loss of generality that R has a superficial element a of degree 1. Let ν be the smallest integer such that $\mathcal{L}(M^n/M^{n+1}) = e$ for **all** $n \geq \nu$. Let $k \geq 0$ be any integer; then by Theorem 12.5 $\mathcal{L}(M^k/M^{k+1})$ $= \mathcal{L}(M^k/M^k a) - \mathcal{L}(M^{k+1}/M^k a) = e - \mathcal{L}(M^{k+1}/M^k a)$. Therefore, $\mathcal{L}(M^k/M^{k+1}) \leq e$ and equality holds if and only if $M^{k+1} = M^k a$. But if $M^{k+1} = M^k a$, then $M^{n+1} = M^n a$ for all $n \geq k$, and hence by Theorem 12.5 $\mathcal{L}(M^n/M^{n+1}) = e$ for all $n \geq k$. In this case $k \geq \nu$, and this proves the proposition.

<u>Definition</u>. Let k_1 be the smallest integer such that $L(R/M^{k_1}) = ek_1 - \rho$; and let k_2 be the smallest integer such that $\Lambda M^{k_2} = M^{k_2}$.

<u>Theorem 12.11</u>. <u>We have</u> $k_1 = k_2 = \nu$, <u>where</u> ν <u>is the integer of Theorem</u> 12.10.

<u>Proof</u>. We have $\mathcal{L}(\Lambda/M^{k_1}) = \mathcal{L}(\Lambda/R) + \mathcal{L}(R/M^{k_1}) = \rho + (ek_1 - \rho)$ $= ek_1$; and we have $\mathcal{L}(\Lambda/\Lambda M^{k_1}) = ek_1$ by Theorem 12.4. Since $M^{k_1} \subset \Lambda M^{k_1}$, we conclude that $M^{k_1} = \Lambda M^{k_1}$. Therefore, $k_2 \leq k_1$.

On the other hand, we have $\mathcal{L}(R/M^{k_2}) = \mathcal{L}(R/\Lambda M^{k_2}) = \mathcal{L}(\Lambda/\Lambda M^{k_2})$ $- \mathcal{L}(\Lambda/R) = ek_2 - \rho$ by Theorem 12.4, and hence $k_1 \leq k_2$.

By Theorem 12.2 there is a **regular element x in Λ such that** $\Lambda x = \Lambda M$. If $n \geq k_2$, then $M^n = \Lambda M^n = \Lambda x^n$, and hence $M^n/M^{n+1} \cong \Lambda/\Lambda x$. Therefore, $\mathcal{L}(M^n/M^{n+1}) = e$ by Theorem 12.4, and thus $k_2 \geq \nu$.

If $k_1 > \nu$, then $\mathscr{L}(M^{k_1-1}/M^{k_1}) = e$, and hence $\mathscr{L}(R/M^{k_1-1})$ $= \mathscr{L}(R/M^{k_1}) - \mathscr{L}(M^{k_1-1}/M^{k_1}) = (ek_1 - \rho) - e = e(k_1 - 1) - \rho$, contradicting the minimality of k_1. Thus $k_1 \leq \nu$ and hence $k_1 = k_2 = \nu$.

Definition. If A is an R-submodule of Q, we define the *inverse of A* by $A^{-1} = \{q \in Q \mid qA \subset R\}$. Since R has a principal M-primary ideal, it is easily seen that $R \subsetneq M^{-1}$. Because R is a local ring, we have $AA^{-1} = R$ if and only if A is isomorphic to R. Thus we see that $MM^{-1} = R$ if and only if R is a discrete valuation ring. If R is not a discrete valuation ring, then $MM^{-1} = M$; and thus M^{-1} is a Noetherian, semi-local, 1-dimensional Cohen-Macaulay ring that properly contains R.

Theorem 12.12. *We have* $M^n(M^n)^{-1} = \Lambda^{-1}$ *for all* $n \geq \nu$.

Proof. By Theorem 12.2, there is an element x in Λ such that $\Lambda x = \Lambda M$. If $n \geq \nu$ and $a = x^n$, we have $\Lambda a = M^n$. Hence $\Lambda^{-1} = a(M^n)^{-1}$ and thus $\Lambda^{-1} = \Lambda\Lambda^{-1} = \Lambda a(M^n)^{-1} = M^n(M^n)^{-1}$.

Theorem 12.13. R *is not a discrete valuation ring if and only if* $M^{-1} \subset \Lambda$.

Proof. Suppose that R is not a discrete valuation ring. Then we have $M^{-1}M = M$. By Theorem 12.1, R has a superficial element a of degree s such that $\Lambda = M^s a^{-1}$. Thus $M^{-1}\Lambda = M^{-1}M^s a^{-1} = M^s a^{-1} = \Lambda$, and hence $M^{-1} \subset \Lambda$.

On the other hand suppose that $M^{-1} \subset \Lambda$. If R is a discrete valuation ring, then $M = Rb$ for some $b \in M$. Therefore, $b^{-1} \in M^{-1} \subset \Lambda$, and thus $\Lambda M = \Lambda$. But this cannot happen since Λ is a finite R-module by Theorem 12.1, and thus R is not a discrete valuation ring.

In Theorem 12.15 we shall find necessary and sufficient conditions for M^{-1} to equal Λ.

Theorem 12.14. $\rho \geq e - 1$.

Proof. We have $\mathcal{L}(\Lambda/M) = \mathcal{L}(\Lambda/R) + \mathcal{L}(R/M) = \rho + 1$ by Theorem 12.4. Since $M \subset \Lambda M$ and $\mathcal{L}(\Lambda/\Lambda M) = e$ by Theorem 12.4, we have $\rho + 1 \geq e$.

The next proposition gives necessary and sufficient conditions for ρ to equal $e - 1$.

Theorem 12.15. The following statements are equivalent:

(1) $\rho = e - 1$.

(2) $\Lambda = M^{-1}$ or R is a discrete valuation ring.

(3) There exists $a \in R$ such that $M^{-1}a = M$.

(4) There exists $a \in R$ such that $M^2 = Ma$.

(5) $\mathcal{L}(M/M^2) = e$ and R has a superficial element of degree 1.

(6) $\mathcal{L}(M^n/M^{n+1}) = e$ for every $n \geq 1$.

Proof. (1) ⟹ (2). Suppose that R is not a discrete valuation ring. Then $M^{-1} \subset \Lambda$ by Theorem 12.12. Now $\mathcal{L}(\Lambda M/M) = \mathcal{L}(\Lambda/M) - \mathcal{L}(\Lambda/\Lambda M) = \mathcal{L}(\Lambda/R) + \mathcal{L}(R/M) - \mathcal{L}(\Lambda/\Lambda M) = \rho + 1 - e = 0$ by Theorem 12.4. Therefore, $\Lambda M = M$ and hence $\Lambda \subset M^{-1}$.

(2) ⟹ (3). If R is a discrete valuation ring, then $M = Rb$ for some $b \in R$ and hence $M^{-1} = R(1/b)$. Letting $a = b^2$ we have $M^{-1}a = Rb = M$. Hence we can assume that $\Lambda = M^{-1}$. Then R is not a discrete valuation ring by Theorem 12.13 and thus $\Lambda M = M^{-1}M = M$. By Theorem 12.2, $\Lambda M = \Lambda x = M$, and hence $x = a \in M$, showing that $M^{-1}a = M$.

(3) ⟹ (4). Since $M^{-1}a = M$, we have $M^2 = M^{-1}Ma$. Thus we can assume that $M^{-1}M \neq M$, and hence $M^{-1}M = R$. But then R is a discrete valuation ring and $M = Rb$ for some $b \in R$. Therefore $M^2 = Rb^2 = Mb$.

(4) ⟹ (5), and (6). Since $M^2 = Ma$, we have $M^{n+1} = M^n a$ for all $n \geq 1$. Hence a is superficial of degree 1, and by Theorem 12.5, $\mathcal{L}(M^n/M^{n+1}) = \mathcal{L}(M^n/M^n a) = e$ for all $n \geq 1$.

(5) ⟹ (4). Let a be a superficial element of degree 1. Then $\mathcal{L}(M/Ma) = e$ by Theorem 12.5. But $Ma \subset M^2$ and $\mathcal{L}(M/M^2) = e$ by assumption. Thus we have $Ma = M^2$.

(6) ==> (1). For large n we have $\mathcal{L}(R/M^n) = en - \rho$. But $\mathcal{L}(R/M^n) = \mathcal{L}(R/M) + \mathcal{L}(M/M^2) + \ldots + \mathcal{L}(M^{n-1}/M^n) = 1 + (n-1)e$ by assumption. Therefore $en - \rho = 1 + (n-1)e = en - (e-1)$, and hence $\rho = e - 1$.

Northcott has proved the following theorem (see [20] and [22]).

Theorem 12.16 (Northcott). *The following statements are equivalent:*

(1) *Every ideal of* R *is principal* (*i.e.*, R *is a discrete valuation ring*).

(2) $P(n) = n$; *where* $P(n)$ *is the Hilbert polynomial of* R.

(3) $\rho = 0$.

(4) $e = 1$.

Proof. The implications (1) ==> (2) ==> (3) are trivial, and (3) ==> (4) follows immediately from Theorem 12.14. Hence assume (4); that is, $e = 1$. Then there is an integer $n > 0$ such that M^n is a principal ideal of R. It is an immediate consequence of this that $MM^{-1} = R$, and hence M is a principal ideal of R. Therefore, R is a discrete valuation ring.

We shall prove an analogous theorem to characterize the situation where every ideal of R can be generated by two elements. However, there is no similar theorem to characterize $\rho = 2$, or the situation where every ideal can be generated by three elements.

Theorem 12.17. *The following statements are equivalent:*

(1) *Every ideal of* R *can be generated by two elements, but* R *is not a discrete valuation ring.*

(2) $P(n) = 2n - 1$, *where* $P(n)$ *is the Hilbert polynomial of* R.

(3) $\rho = 1$.

(4) $e = 2$.

If any of these conditions hold, then R *has a superficial element of degree* 1.

Proof. (1) $\Longrightarrow$ (2). If M^n is principal for any $n \geq 1$, then $M^{-1}M = R$ and hence R is a discrete valuation ring contrary to assumption. Thus $\mathcal{L}(M^n/M^{n+1}) \geq 2$ for all $n \geq 1$ because $\mathcal{L}(M^n/M^{n+1})$ is the minimal number of generators required to generate M^n. By assumption $\mathcal{L}(M^n/M^{n+1}) \leq 2$ for all $n \geq 1$. Hence $\mathcal{L}(M^n/M^{n+1}) = 2$ for all $n \geq 1$. Therefore, $e = 2$, and by Theorem 12.15, we have $\rho = e - 1 = 2 - 1 = 1$. Thus $\underline{P}(n) = 2n - 1$.

(2) $\Longrightarrow$ (3). Trivial.

(3) $\Longrightarrow$ (4). By Theorem 12.14, we have $e \leq \rho + 1 = 2$; and by Theorem 12.16, we have $e > 1$. Therefore, $e = 2$.

(4) $\Longrightarrow$ (1). By Theorem 12.10, $\mathcal{L}(M^{\nu-1}/M^{\nu}) < e = 2$, and thus $\mathcal{L}(M^{\nu-1}/M^{\nu}) = 1$. Hence $M^{\nu-1}$ is a principal ideal of R and the same is true of $(M^{\nu-1})^n$ for all $n \geq 1$. If $\nu - 1 > 0$, then $1 = \mathcal{L}(M^{(\nu-1)n}/M^{(\nu-1)n+1}) = e$ for some large n. But $e = 2$, and this contradiction shows that $\nu - 1 = 0$. Hence by Theorem 12.10 we have $\mathcal{L}(M^n/M^{n+1}) = e = 2$ for all $n \geq \nu = 1$. This is one of the equivalent conditions of Theorem 12.15, and we conclude that R has a superficial element a of degree 1 such that $M^2 = Ma$. Therefore by Theorem 12.8 every M-primary ideal can be generated by $e = 2$ elements.

Suppose that P is a nonzero rank 0 prime ideal of R. If $P \subset Ra$, then $a \notin P$ since a is a regular element of R, and hence $P = Pa$. But then $P = Pa^n \subset M^{n+1}$ for all $n \geq 1$, and thus $P = 0$. This contradiction shows that $P \not\subset Ra$. Choose $b \in P$ such that $b \notin Ra$. If a and b are not linearly independent modulo M^2, then $b \in Ra + M^2 = Ra$. This contradiction, and the fact that M can be generated by two elements shows that $M = (a,b)$. Therefore, R/Rb is a 1-dimensional Noetherian local ring whose maximal ideal is principal and thus R/Rb is a discrete valuation ring. Hence Rb is a prime ideal; and since $Rb \subset P$ and rank $P = 0$, we see that $P = Rb$. We have shown that every rank 0 prime ideal of R is a principal ideal.

Suppose that I is a P-primary ideal of R. Since $P^n \subset I$ for some

$n \geq 1$, there is a smallest integer $t \geq 1$ such that $b^t \in I$. But then $I = Rb^t$. For if not, choose $y \in I$ such that $y \notin Rb^t$. Then $y = rb^k$ where $k < t$ and $r \notin Rb = P$. But I is P-primary and $r \notin P$ implies that $b^k \in I$. This contradiction shows that $I = Rb^t$. Therefore every P-primary ideal is a power of Rb.

It is easily seen that the intersection of a finite number of ideals that are powers of principal prime ideals is again a principal ideal. Thus every unmixed rank 0 ideal of R is a principal ideal.

Now let J be a nonzero ideal of R that is not M-primary and is not unmixed of rank 0. Then $J = A \cap I$, where A is an M-primary ideal and I is unmixed of rank 0. Then $I = Rc$ for some $c \in M$ and hence $J = Bc$, where B is an ideal of R. If B is M-primary, or unmixed of rank 0, then B can be generated by two elements, or 1 element, respectively. But then in all cases $J = Bc$ can be generated by two elements. Thus we can assume that B is neither M-primary nor unmixed of rank 0. As in the case of J we see that $B = B_1c_1$, where $c_1 \in M$ and B_1 is an ideal of R. Therefore, $J = B_1c_1c \subset M^2$. If this process were to continus indefinitely, we would have $J \subset M^n$ for all $n \geq 0$, and hence $J = 0$. This contradiction shows that the process must stop somewhere, and when it does we see that J can be generated by 1 or 2 elements. Therefore, every ideal of R can be generated by two elements.

CHAPTER XIII

GORENSTEIN RINGS

Throughout this chapter R will be a 1-dimensional local Cohen-Macaulay ring. We shall freely use the notation of Chapter XII without further definition.

Definition. R is said to be a Gorenstein ring if the injective dimension of R as an R-module is 1.

The following theorem may be found in [1] and [14] in a more general setting.

Theorem 13.1. *The following statements are equivalent:*

(1) R *is a Gorenstein ring.*

(2) Q *and* K *are injective R-modules.*

(3) K *is an injective R-module.*

(4) $K \cong E(R/M)$, *the injective envelope of* R/M.

(5) $M^{-1}/R \cong R/M$.

(6) M^{-1} *can be generated by two elements.*

(7) $I = I^{-1-1}$ *for every regular ideal of* R.

(8) $\mathcal{L}(I/J) = \mathcal{L}(J^{-1}/I^{-1})$ *for all regular ideals* $J \subset I$ *of* R.

(9) R *has a proper, irreducible, regular, principal ideal.*

Proof. If $p_1,\ldots,p_t$ are the prime ideals of rank 0 in R, then $Q = R_{p_1} \oplus \ldots \oplus R_{p_t}$, and hence Q is a direct sum of Artinian local rings. Let $E = E(R/M)$ and $E_{p_i} = E(R/p_i)$ be the injective envelopes of R/M and R/p_i, respectively. By Theorem 4.5 these are the only indecomposable injective R-modules; the E_{p_i}'s are the only indecomposable injective Q-modules; and every injective R (or Q) module is a direct sum of indecomposable injectives. It follows that $\text{inj. dim}_R Q \leq \text{inj. dim}_Q Q$. On the other hand we have $Q \otimes_R E = 0$ and $Q \otimes_R E_{p_i} \cong E_{p_i}$. Thus if C is an injective R-module, then $Q \otimes_R C$ is an injective Q-module. Hence if we apply the exact functor $Q \otimes_R \cdot$

to an injective resolution for either R or Q over R we obtain an injective resolution for Q as a Q-module. Thus we have proved the following:

inj $\dim_R Q$ = inj.$\dim_Q Q \leq$ inj.$\dim_R R$.

(1) $\Longrightarrow$ (2). By the preceding remarks we have inj.$\dim_Q Q \leq 1$. It is sufficient to prove that Q is an injective Q-module. For then by the preceding equation Q would be an injective R-module. Since Q is a direct sum of Artinian local rings, it is sufficient to prove that if S is an Artinian local ring such that inj.$\dim._S S \leq 1$, then S is an injective S-module.

Let N be the maximal ideal of S. Then there is a nonzero element x in S such that Nx = 0, and hence Sx = S/N. We have $\mathrm{Ext}_S^1(S/N,S) \cong \mathrm{Ext}_S^1(Sx,S) \cong \mathrm{Ext}_S^2(S/Sx,S) = 0$. By induction on length we see that $\mathrm{Ext}_S^1(A,S) = 0$ for every S-module A of finite length. Because every ideal of S has finite length, we have shown that S is an injective S-module.

(2) $\Longrightarrow$ (3). This is a trivial assertion.

(3) $\Longrightarrow$ (4). It is easy to see that $M^{-1}/R \neq 0$ and that K is an essential extension of M^{-1}/R. Since M^{-1}/R is isomorphic to a finite direct sum of copies of R/M, it follows that K is isomorphic to a finite direct sum of copies of E(R/M). But K is an indecomposable R-module by Theorem 4.7, and hence K is isomorphic to E(R/M).

(4) $\Longrightarrow$ (1). It is sufficient to prove that Q is an injective R-module. From the given exact sequence $0 \to R \to Q \to E \to 0$, we obtain the exact sequence:

$$(*) \qquad 0 \to \mathrm{Hom}_R(E,E) \to \mathrm{Hom}_R(Q,E) \to E \to 0.$$

Now $\mathrm{Hom}_R(E,E) \cong H$, the completion of R, by Theorem 4.5, and hence $\mathrm{Hom}_R(E,E)$ is a flat R-module. Using the abbreviation w.dim. for weak dimension, we observe that $\mathrm{w.dim}_R E = 1$ since $E \cong K$, and Q is a flat R-module. Thus from exact sequence (*) we see that

$w.\dim_R \mathrm{Hom}_R(Q,E) \leq 1$.

If I is an ideal of R, then by [3, Ch. VI, Prop. 5.3], we have:

$$\mathrm{Hom}_R(\mathrm{Ext}_R^n(R/I,Q),E) \cong \mathrm{Tor}_n^R(R/I,\mathrm{Hom}_R(Q,E))$$

for all $n \geq 0$. Hence we have $\mathrm{Ext}_R^n(R/I,Q) = 0$ for all $n \geq 2$. Thus $\mathrm{inj.dim}_R Q \leq 1$. By the equations at the beginning of the proof we have $\mathrm{inj.dim}_Q Q \leq 1$. As in the proof of (1) ==> (2), we see that Q is an injective R-module. Therefore, $\mathrm{inj.dim}_R R = 1$, and R is a Gorenstein ring.

(4) ==> (5). Since $M^{-1}/R = \{x \in K \mid Mx = 0\}$ and $R/M \cong \{x \in E \mid Mx = 0\}$, we see that $M^{-1}/R \cong R/M$.

(5) ==> (6). This is an obvious assertion.

(6) ==> (5). If $M^{-1}M = R$, then M^{-1} is cyclic and hence $M^{-1}/R \cong R/M$. Hence we can assume that $M^{-1}M = M$. But then $\mathcal{L}(M^{-1}/M) = \mathcal{L}(M^{-1}/MM^{-1}) = 2$. Therefore $\mathcal{L}(M^{-1}/R) = \mathcal{L}(M^{-1}/M) - \mathcal{L}(R/M) = 1$, and we see that $M^{-1}/R \cong R/M$.

(5) ==> (7). Let I be a regular ideal of R and let a be a regular element of I. Then multiplication by a annihilates R/I. On the other hand multiplication by a is an isomorphism on Q. Hence $\mathrm{Ext}_R^n(R/I,Q) = 0$ for all $n \geq 0$. As a consequence we have $\mathrm{Ext}_R^1(R/I,R) \cong \mathrm{Hom}_R(R/I,K) \cong \mathrm{Ann}_K I = I^{-1}/R$. Another consequence is that every R-homomorphism from I into Q can be extended to an R-homomorphism of R into Q. This latter statement implies directly that there is a canonical isomorphism of I^{-1} and $\mathrm{Hom}_R(I,R)$.

Since Ra is an M-primary ideal we have a composition series of ideals:

$$Ra = I_0 \subset I_1 \subset \ldots \subset I_n = I$$

such that $I_{j+1}/I_j \cong R/M$. Because I_0 is a principal ideal, we have $I_0 = I_0^{-1-1}$. We will assume that $I_j = I_j^{-1-1}$ and prove that $I_{j+1} = I_{j+1}^{-1-1}$. By induction this will establish (7).

We have an exact sequence:

$$0 \to \mathrm{Hom}_R(I_{j+1},R) \to \mathrm{Hom}_R(I_j,R) \to \mathrm{Ext}^1_R(R/M,R).$$

By our preceding remarks we have $\mathrm{Ext}^1_R(R/M,R) \cong M^{-1}/R$, and by assumption $M^{-1}/R \cong R/M$. Thus the exact sequence can be written in the form:

$$0 \to I^{-1}_{j+1} \xrightarrow{\alpha} I^{-1}_j \xrightarrow{\beta} R/M$$

where α is the inclusion map. If $\beta = 0$, then $I^{-1}_{j+1} = I^{-1}_j$, and hence $I_{j+1} \subset I^{-1-1}_{j+1} = I^{-1-1}_j = I_j$. But this is a contradiction, and thus $\beta \neq 0$, and so β is onto. We can now produce from this sequence another exact sequence:

$$0 \to I^{-1-1}_j \xrightarrow{\gamma} I^{-1-1}_{j+1} \xrightarrow{\delta} R/M$$

where γ is the inclusion map. If $\delta = 0$, then $I_j = I^{-1}_{j+1}$ and we have a contradiction. Hence δ is onto. Since $I^{-1-1}_j = I_j$, we have proved that $I^{-1-1}_{j+1}/I_j \cong R/M$. Now I_{j+1}/I_j is a nonzero submodule of I^{-1-1}_{j+1}/I_j, and hence $I_{j+1} = I^{-1-1}_{j+1}$.

(7) ⟹ (8). Let $J \subset I$ be regular ideals of R. They are M-primary ideals of R and hence we can find a composition series of ideals of R:

$$J = I_0 \subset I_1 \subset \ldots \subset I_n = I$$

such that $I_{j+1}/I_j \cong R/M$. Then we have a series:

$$I^{-1} = I^{-1}_n \subset I^{-1}_{n-1} \subset \ldots \subset I^{-1}_0 = J^{-1},$$

and $I^{-1}_{j+1} \neq I^{-1}_j$. Thus $\mathcal{L}(I/J) \leq \mathcal{L}(J^{-1}/I^{-1})$. By the same argument we have $\mathcal{L}(J^{-1}/I^{-1}) \leq \mathcal{L}(I^{-1-1}/J^{-1-1})$. Since $I = I^{-1-1}$ and $J = J^{-1-1}$, we see that $\mathcal{L}(I/J) = \mathcal{L}(J^{-1}/I^{-1})$.

(8) ⟹ (4). By assumption $\mathcal{L}(R/M) = \mathcal{L}(M^{-1}/R)$, and thus $M^{-1}/R \cong R/M$. Since K is an essential extension of M^{-1}/R, we have $K \subset E = E(R/M)$. Let I be a regular ideal of R. Then

$\operatorname{Ann}_K(I) \subset \operatorname{Ann}_E(I)$. We have $\operatorname{Ann}_K(I) = I^{-1}/R$ and $\operatorname{Ann}_E(I) \cong \operatorname{Hom}_R(R/I,E)$. Since $\operatorname{Hom}_R(R/M,E) \cong R/M$ it is easy to see by induction on the length of a module that $\mathcal{L}(\operatorname{Hom}_R(R/I,E)) = \mathcal{L}(R/I)$. By assumption we have $\mathcal{L}(I^{-1}/R) = \mathcal{L}(R/I)$. Therefore, we have $\operatorname{Ann}_K(I) = \operatorname{Ann}_E(I)$. By Theorem 4.5 we see that $E = \bigcup_n \operatorname{Ann}_E(M^n)$. This shows that $K = E$.

(4) $\Longrightarrow$ (9). Let a be a regular element of M and let $x = a^{-1} + R$ in K. Then $\operatorname{Ann}_R(x) = Ra$. But since $K \cong E$, $\operatorname{Ann}_R(x)$ is an irreducible ideal of R by Theorem 4.5.

(9) $\Longrightarrow$ (5). Let a be a regular element of M such that Ra is an irreducible ideal of R. Let $\mathcal{S}$ be the multiplicatively closed set in R consisting of the powers of a. Then clearly $R_{\mathcal{S}} = Q$. Since Ra is an irreducible M-primary ideal, there exists an element $x \in E$ such that $\operatorname{Ann}_R(x) = Ra$ (by Theorem 4.5). Since E is a divisible R-module we have $E = aE$. Hence we can find a sequence of elements $\{x_n\}$ in E such that $x_1 = x$ and $ax_n = x_{n-1}$ for $n > 1$. We define $f : K \to E$ by $f(b/a^n + R) = bx_n$. It is easy to verify that f is a well defined monomorphism. Hence we have $M^{-1}/R \cong R/M$.

Remarks. If R is an integral domain, then (1), (2), and (3) of the theorem are trivially equivalent.

The following theorem may be found in [1] and [14].

Theorem 13.2. *If the maximal ideal M of R can be generated by two elements, then R is a Gorenstein ring.*

Proof. Let $p_1,\ldots,p_t$ be the prime ideals of rank 0 of R. Then $\bigcup_{i=1}^{t} p_i$ is the set of zero divisors in R. If $M - M^2 \subset \bigcup_{i=1}^{t} p_i$, then $M \subset p_i$ for some i. This contradiction shows that there is an element $a \in M - M^2$ that is not a zero divisor in R. We can find an element $b \in M$ such that $M = (a,b)$. Let $\bar{R} = R/Ra$; since Ra is an M-primary ideal of R, $\bar{R}$ is an Artinian local ring whose maximal ideal is generated by a single element. Thus every ideal of $\bar{R}$ is equal to some power of the maximal ideal of $\overline{R}$. Hence the 0-ideal of $\overline{R}$ is an irreducible ideal, and thus Ra is an irreducible ideal of R. Therefore by Theorem 13.1, R is a Gorenstein ring.

We let Λ denote the first neighborhood ring of R as in Chapter XII.

Theorem 13.3. *Every ideal of R can be generated by two elements if and only if R is a Gorenstein ring and $\Lambda = M^{-1}$ or $\Lambda = R$.*

Proof. By Theorem 12.1, R is a discrete valuation ring if and only if $\Lambda = R$. Thus we may assume that R is not a discrete valuation ring. Suppose that every ideal of R can be generated by two elements. Then R is a Gorenstein ring by Theorem 13.2. By Theorem 12.17 we have $\rho = e - 1$, and hence $\Lambda = M^{-1}$ by Theorem 12.15.

Conversely, suppose that R is a Gorenstein ring and that $\Lambda = M^{-1}$. Since $\mathcal{L}(M^{-1}/R) = 1$ by Theorem 13.1, and $\mathcal{L}(\Lambda/R) = \rho$ by Theorem 12.4, we have $\rho = 1$. Thus by Theorem 12.17, every ideal of R can be generated by two elements.

Theorem 13.4. *Assume that R is a Gorenstein ring. Then $\rho \leq \frac{1}{2}\nu e$ and $\rho = \frac{1}{2}\nu e$ if and only if $\Lambda^{-1} = M^{\nu}$.*

Proof. Since $\Lambda M^{\nu} = M^{\nu}$, we have $M^{\nu} \subset \Lambda^{-1}$. Thus $\rho = \mathcal{L}(\Lambda/R)$ $= \mathcal{L}(R/\Lambda^{-1}) \leq \mathcal{L}(R/M^{\nu}) = e\nu - \rho$, from which it follows that $\rho \leq \frac{1}{2}\nu e$ and $M^{\nu} = \Lambda^{-1}$ if and only if $\rho = \frac{1}{2}\nu e$.

Northcott has given sufficient conditions for the equation $\Lambda^{-1} = M^{e-1}$ and for the equation $\rho = \frac{1}{2}e(e-1)$ ([20, Lemma 4] and [23, Th. 1 and 2]). The next theorem provides a link between these two equations.

Theorem 13.5. *Assume that R is a Gorenstein ring. Then the following two statements are equivalent:*

(1) $\nu = e - 1$ *and* $\rho = \frac{1}{2}e(e-1)$.

(2) $\Lambda^{-1} = M^{e-1}$.

Proof. (1) $\Longrightarrow$ (2). We have $\rho = \frac{1}{2}(e-1)e = \frac{1}{2}\nu e$, and thus by Theorem 13.4, $\Lambda^{-1} = M^{\nu} = M^{e-1}$.

(2) $\Longrightarrow$ (1). Since $\Lambda M^{\nu} = M^{\nu}$, we have $M^{\nu} \subset \Lambda^{-1} = M^{e-1}$, and thus $e - 1 \leq \nu$. On the other hand $\Lambda M^{e-1} = \Lambda\Lambda^{-1} = \Lambda^{-1} = M^{e-1}$, and thus $\nu \leq e - 1$ by the minimality of ν. Therefore $\nu = e - 1$, and $\Lambda^{-1} = M^{\nu}$. Thus by Theorem 13.4, $\rho = \frac{1}{2}\nu e = \frac{1}{2}(e-1)e$.

Lemma 13.6. *Assume that the maximal ideal* M *of* R *can be generated by two elements. If* $\mathcal{L}(M^q/M^{q+1}) < q + 1$, *then* $\mathcal{L}(M^n/M^{n+1}) \leq \mathcal{L}(M^m/M^{m+1})$ *for every* n *and* m *such that* $q \leq m \leq n$.

Proof. Because of an obvious recursive argument it is sufficient to prove that $\mathcal{L}(M^{q+1}/M^{q+2}) \leq \mathcal{L}(M^q/M^{q+1})$. Let a and b be elements of R that generate M, and define F_i to be the subspace of M^q/M^{q+1} generated by the images of $b^q, ab^{q-1}, a^2b^{q-2}, \ldots, a^ib^{q-i}$ in M^q/M^{q+1}. Similarly define G_j to be the subspace of M^{q+1}/M^{q+2} generated by the images of $b^{q+1}, ab^q, a^2b^{q-1}, \ldots, a^jb^{q+1-j}$ in M^{q+1}/M^{q+2}.

We will prove first that if $F_k = F_{k-1}$ (for $k > 0$), then $G_{k+1} = G_{k-1}$. For if $F_k = F_{k-1}$, then we have:

$$(1) \quad a^kb^{q-k} \in (b^q, ab^{q-1}, \ldots, a^{k-1}b^{q-k+1}) + M^{q+1}.$$

Multiplying (1) by a and then by b we obtain:

$$(2) \quad a^{k+1}b^{q-k} \in (ab^q, a^2b^{q-1}, \ldots, a^kb^{q-k+1}) + M^{q+2}$$

and

$$(3) \quad a^kb^{q-k+1} \in (b^{q+1}, ab^q, \ldots, a^{k-1}b^{q-k+2}) + M^{q+2}.$$

Substituting (3) in (2) we see that both $a^{k+1}b^{q-k}$ and a^kb^{q-k+1} are in $(b^{q+1}, ab^q, \ldots, a^{k-1}b^{q-k+2}) + M^{q+2}$. Therefore, we have shown that $G_{k+1} = G_{k-1}$.

By hypothesis there is an integer p such that $\mathcal{L}(F_p) = p$ and $\mathcal{L}(F_s) = s + 1$ if $0 \leq s < p$. If $p = 0$, then $b^q \in M^{q+1}$ and hence

b^{q+1} and ab^q are in M^{q+2}. Therefore $G_1 = 0$, and we have $\mathcal{L}(G_{p+1}) \leq \mathcal{L}(F_p)$. If $p > 0$, then $F_p = F_{p-1}$, and hence by the preceding paragraph $G_{p+1} = G_{p-1}$. Thus $\mathcal{L}(G_{p+1}) \leq \mathcal{L}(F_p)$ in this case also. Therefore, there exists a largest integer t such that $\mathcal{L}(G_{t+1}) \leq \mathcal{L}(F_t)$. We will suppose that $t < q$ and arrive at a contradiction.

Case I: $\mathcal{L}(F_{t+1}) = \mathcal{L}(F_t) + 1$.

We then have $\mathcal{L}(G_{t+2}) \leq \mathcal{L}(G_{t+1}) + 1 \leq \mathcal{L}(F_t) + 1 = \mathcal{L}(F_{t+1})$. This contradicts the maximality of t and hence this case cannot arise.

Case II. $\mathcal{L}(F_{t+1}) = \mathcal{L}(F_t)$.

In this case we have $F_{t+1} = F_t$, and hence by the preceding paragraph $G_{t+2} = G_t$. Thus $\mathcal{L}(G_{t+2}) = \mathcal{L}(G_t) \leq \mathcal{L}(G_{t+1}) \leq \mathcal{L}(F_t) = \mathcal{L}(F_{t+1})$. This also contradicts the maximality of t. Since these are the only two cases that can arise, we have $t = q$ and the lemma is proved.

Theorem 13.7. (Step Theorem). Assume that the maximal ideal M of R can be generated by two elements. Then

$$\mathcal{L}(M^n/M^{n+1}) = \begin{cases} n+1 & \text{if } n \leq e-1 \\ e & \text{if } n \geq e-1 \end{cases}.$$

Proof. Since there are only n + 1 monomials of degree n in the two generators of M, we have $\mathcal{L}(M^n/M^{n+1}) \leq n + 1$ for all $n \geq 0$. Since $\mathcal{L}(M^n/M^{n+1})$ is bounded we conclude from Lemma 13.6 that there is an integer $k > 0$ such that $\mathcal{L}(M^n/M^{n+1}) = n + 1$ for $n \leq k$ and $\mathcal{L}(M^n/M^{n+1})$ is a nonincreasing function for $n \geq k$. By Theorem 12.10, we have $\nu = k$ and hence $e = \mathcal{L}(M^\nu/M^{\nu+1}) = k + 1$. Thus $\nu = e - 1$ and the theorem follows immediately from Theorem 12.10.

Corollary 13.8. Assume that the maximal ideal M of R can be generated by two elements. Then:

(1) $\nu = e - 1$.

(2) $\rho = \frac{1}{2}(e-1)e$.

(3) $\Lambda^{-1} = M^{e-1}$ <u>and</u> $\Lambda = (M^{e-1})^{-1}$.

<u>Proof</u>. It is an immediate consequence of Theorem 12.10 and 13.7 that $\nu = e - 1$. Using Theorem 13.7 we have $e\nu - \rho = \mathcal{L}(R/M^{\nu})$ $= \mathcal{L}(R/M) + \mathcal{L}(M/M^2) + \ldots + \mathcal{L}(M^{\nu-1}/M^{\nu}) = 1 + 2 + \ldots + \nu$ $= \frac{1}{2}\nu(\nu+1)$. Therefore, $\rho = e\nu - \frac{1}{2}\nu(\nu + 1) = e(e-1) - \frac{1}{2}(e-1)e$ $= \frac{1}{2}(e-1)e$.

Since the maximal ideal of R can be generated by two elements, R is a Gorenstein ring by Theorem 13.2. Therefore, $\Lambda^{-1} = M^{e-1}$ by Theorem 13.5 and $\Lambda = \Lambda^{-1-1} = (M^{e-1})^{-1}$.

<u>Remarks</u>. Theorem 13.7 and (2) and (3) of Corollary 13.8 were proved by Northcott in the special case where R is a homomorphic image of a 2-dimensional regular local ring [23, Th. 1 and Th. 2 and (2.3)]. One consequence of the greater generality of our theorems is that the proofs are correspondingly simpler and more elementary in nature.

The next two propositions give useful criteria for determining the multiplicity of R and for deciding whether or not a given element is superficial of degree 1.

<u>Theorem 13.9</u>. <u>Assume that the maximal ideal</u> M <u>of</u> R <u>can be generated by two elements</u>. <u>If</u> $a \in M - M^2$, <u>let</u> n <u>be the smallest integer such that</u> $M^n \subset Ra$. <u>Then</u> $n \geq e$ <u>and</u> $n = e$ <u>if and only if</u> a <u>is superficial of degree</u> 1.

<u>Proof</u>. We can assume that there is an integer n such that $M^n \subset Ra$, for otherwise there is nothing to prove. Because $a \in M - M^2$, we can choose $b \in M - M^2$ such that $M = (a,b)$. If q is the largest integer such that $b^n \in M^q a$, then:

(1) Either $q \geq n - 1$ or $n > e$.

Suppose that (1) is false. Then $q < n - 1$ and $n \leq e$, and thus $q + 1 < n \leq e$. Now b^n is a linear combination of monomials of degree $q + 1$ because b^n is in $M^q a$. Since $b^n \notin M^{q+1}a$, the coefficient of one of these monomials is a unit in R. Hence we can solve the equation for this monomial and express it as a linear combination of the other monomials of degree $q + 1$ and b^n. Since $n > q + 1$, this shows that $\mathcal{L}(M^{q+1}/M^{q+2}) < q + 2$. Hence by Theorem 13.7, $q + 1 > e - 1$. This contradicts $q + 1 < e$ and establishes (1).

Assume that a is superficial of degree 1. By Theorem 12.5 we have $\mathcal{L}(M^\nu/M^\nu a) = e$. Since $\mathcal{L}(M^\nu/M^{\nu+1}) = e$ by Theorem 12.10 and $M^\nu a \subset M^{\nu+1}$, we see that $M^\nu a = M^{\nu+1}$. Hence by the minimality of n we have $n \leq \nu + 1$. By Corollary 13.8, $\nu + 1 = e$ and thus $n \leq e$. It follows from (1) that $q \geq n - 1$. Therefore $b^n \in M^q a \subset M^{n-1}a$ and hence $M^n = M^{n-1}a$. We conclude from Theorem 12.5 that $\mathcal{L}(M^{n-1}/M^n) = e$ and hence by Theorem 13.7 we have $n - 1 \geq e - 1$; that is $n \geq e$. Thus we have $n = e$ in this case.

Now assume that a is not superficial of degree 1. If $n \leq e$, then $q \geq n - 1$ by (1). Hence $b^n \in M^q a \subset M^{n-1}a$, showing that $M^n = M^{n-1}a$. But then a is superficial of degree 1 contrary to assumption. Thus $n > e$ in this case.

<u>Theorem 13.10.</u> <u>Assume that the maximal ideal</u> M <u>of</u> R <u>can be generated by two elements, and let</u> $a \in M$. <u>Then</u> a <u>is superficial of degree</u> 1 <u>if and only if</u> $M^{e-1}a = M^e$. <u>Furtheremore</u>, e <u>is the smallest integer with this property</u>.

<u>Proof.</u> Of course if $M^{e-1}a = M^e$, then a is superficial of degree 1 by definition. Conversely, assume that a is superficial of degree 1. If $M^{k-1}a = M^k$, then $\mathcal{L}(M^{k-1}/M^k) = e$ by Theorem 12.5. Thus $k - 1 \geq \nu$ by Theorem 12.10. On the other hand $\mathcal{L}(M^\nu/M^{\nu+1}) = e$ by Theorem 12.10 and $\mathcal{L}(M^\nu/M^\nu a) = e$ by Theorem 12.5. Since $M^\nu a \subset M^{\nu+1}$, we see that $M^\nu a = M^{\nu+1}$. But $\nu = e - 1$ by Corollary 13.8 and thus $M^{e-1}a = M^e$.

CHAPTER XIV
MULTIPLICITIES

Throughout this chapter R will be a 1-dimensional, local Cohen-Macaulay ring. As in earlier chapters M is the maximal ideal of R and H is its completion. If A is an Artinian R-module, then L(A) denotes its "divisible" length. If A has finite length in the classical sense, then L(A) = 0, and we denote its classical length by $\mathcal{L}(A)$. We shall use all of the notation of earlier chapters without further explanation.

Lemma 14.1. *Let P be a prime ideal of rank 0 in H and $I \subsetneq J$ P-primary ideals of H such that there does not exist a P-primary ideal properly between I and J. Let N be an unmixed ideal of rank 0 in H such that P does not belong to N. Then $(N \cap J)/(N \cap I)$ is isomorphic to a nonzero ideal of H/P.*

Proof. Clearly $H_P I \subset H_P J$ are PH_P-primary ideals and there are no PH_P-primary ideals properly between them. Since H_P is a Noetherian local ring with maximal ideal PH_P, this means that $H_P J/H_P I$ is a simple H_P-module. Thus $H_P J/H_P I \cong H_P/PH_P$, the quotient field of H/P.

Now multiplications by elements of H - P on J/I are monomorphisms, and thus we have an imbedding $J/I \subset H_P J/H_P I$. This demonstrates that J/I is a torsion-free H/P-module of rank 1. Since J/I is a finitely generated H-module, we conclude that J/I is isomorphic to a nonzero ideal of H/P.

Now $(N \cap J)/(N \cap I) \cong ((N \cap J) + I)/I \subset J/I$, and thus to conclude the proof it is merely necessary to show that $N \cap J \neq N \cap I$. But this follows immediately from comparing a normal decomposition of $N \cap J$ with a normal decomposition of $N \cap I$.

Definition. Let P be a prime ideal of rank 0 in H and let N be the P-primary component of 0 in H. Then it is well known [18, Ch.

III, Th. 5] that there exists a so-called "composition series" for N; that is, a chain of P-primary ideals:

$$N = P_1 \subsetneq P_2 \subsetneq \cdots \subsetneq P_k = P$$

such that there are no P-primary ideals properly between P_i and P_{i+1} for all i. Then the <u>latent multiplicity of P</u> is defined by $\mu(P) = k$, the number of terms in the chain. It is also well known (see [18, Ch. III, Th. 5]) that by a Jordan-Holder type of argument the integer $\mu(P)$ is independent of the chain chosen.

<u>Theorem 14.2.</u> <u>Let $P_1,\ldots,P_n$ be the set of prime ideals of rank 0 in H. Then:</u>

$$L(K) = \sum_i \mu(P_i),$$

<u>and $\mu(P_i)$ is the number of times (i.e., the multiplicity) that a simple divisible module corresponding to P_i occurs as a factor module in a composition series of divisible submodules of K.</u>

<u>Proof.</u> Let $0 = N_1 \cap \ldots \cap N_n$ be a normal decomposition of 0 in H, where N_i is P_i-primary. By patching together "composition series" for each N_i we obtain a descending chain of unmixed ideals of rank 0 in H given by:

$$(1)\quad H \supsetneq P_1 \supsetneq P_{12} \supsetneq \cdots \supsetneq N_1 \supsetneq N_1 \cap P_2 \supsetneq \cdots \supsetneq N_1 \cap N_2$$
$$\supsetneq \cdots \supsetneq \bigcap_i N_i = 0.$$

If we tensor this chain with K, we obtain by Theorem 6.6 a descending chain of divisible submodules of K:

$$(2)\quad K \supsetneq K \otimes_R P_1 \supsetneq \cdots \supsetneq K \otimes_R N_1 \supsetneq \cdots \supsetneq K \otimes_R (N_1 \cap N_2)$$
$$\supsetneq \cdots \supsetneq K \otimes_R \bigcap_i N_i = 0.$$

Letting $N \cap P_{i,t} \supsetneq N \cap P_{i,t+1}$ be a typical step in chain (1), we observe that $(N \cap P_{i,t})/(N \cap P_{i,t+1})$ is isomorphic to a nonzero ideal of H/P_i by Lemma 14.1. Thus $(K \otimes_R (N \cap P_{i,t}))/(K \otimes_R (N\cap P_{i,t+1}))$ $\cong K \otimes_R (N \cap P_{i,t})/(N \cap P_{i,t+1})$ is a simple divisible R-module corresponding to P_i by Theorem 8.5. Thus chain (2) is a composition series of divisible submodules of K. Clearly this proves the theorem.

<u>Corollary 14.3</u>. <u>If S is the integral closure of R in Q, then</u>

$$L(S/R) = \sum_{i=1}^{n} (\mu(P_i) - 1).$$

<u>Proof</u>. By Theorem 10.1 and Theorem 14.2 we have $L(S/R) = L(K) - n = \sum_{i=1}^{n} \mu(P_i) - n = \sum_{i=1}^{n} (\mu(P_i) - 1)$.

<u>Remarks</u>. Corollary 14.3 gives an immediate proof of the theorem [19, Th. 8] that R is analytically unramified if and only if $\mu(P_i) = 1$ for all $i = 1,\ldots,n$. (See also Theorem 10.2).

<u>Definition</u>. If T is a 1-dimensional, local, Cohen-Macaulay ring, we shall denote its multiplicity by e_T and its reduction number by ρ_T. If T = R, we shall drop the subscripts.

<u>Remarks</u>. Let B be a strongly unramified extension of R in Q. Then by Theorem 3.6, BM is a regular, maximal ideal of B and BM contains every regular ideal of B. If I is a regular ideal of B, then I is a finitely generated ideal of B, $I = B(I \cap R)$, and $B/I \cong R/(I \cap R)$. $I \cap R$ is a regular (hence M-primary) ideal of R. It follows readily that there is an integer $n > 0$ such that $(BM)^n \subset I$, and thus I is a BM-primary ideal of B. Therefore B/I has finite length over B (and the same finite length over R). In general, a finitely generated torsion B-module is also a finitely generated torsion R-module and has the same finite length over B as over R.

We shall recapitulate some of the results of Theorem 2.11 and other theorems. Let $\mathcal{Q}$ be the divisible submodule of B, and let

$\overline{B} = B/\mathcal{D}$, $\overline{Q} = Q/\mathcal{D}$, and $\overline{R} = R/(\mathcal{D} \cap R)$. Now $\mathcal{D}$ is an ideal of Q, and $\overline{Q}$ is the full ring of quotients of both $\overline{R}$ and $\overline{B}$. $\overline{B}$ is a strongly unramified ring extension of $\overline{R}$ in $\overline{Q}$, and $\overline{B}$ is a reduced $\overline{R}$-module. $\overline{B}$ is a quasi-local ring by Theorem 3.6. Clearly we have $\overline{Q}/\overline{B} \cong Q/B$. Let $\overline{H}$ be the completion of B in the B-topology. Then $\overline{H} = \mathrm{Hom}_B(Q/B,Q/B) \cong \mathrm{Hom}_{\overline{B}}(\overline{Q}/\overline{B},\overline{Q}/\overline{B})$, and hence $\overline{H}$ is also the completion of $\overline{B}$ in the $\overline{B}$-topology. By Theorem 6.6, $\overline{H} \cong H/J$, where $J = \mathrm{Hom}_R(K,B/R)$, and thus $\overline{H}$ is a complete, local, 1-dimensional Cohen-Macaulay ring. If B is a closed component of R, then by Theorem 7.2, $\overline{B}$ is a local, 1-dimensional, analytically irreducible, Cohen-Macaulay ring.

$\mathcal{D}$ is contained in every regular ideal of B, and in fact $\mathcal{D}$ is the intersection of the regular ideals of B. If I is a regular ideal of B, and $\overline{I} = I/$, then $\overline{I}$ is a regular ideal of $\overline{B}$ and every regular ideal of $\overline{B}$ arises in this way. The completion of I over B is equal to the completion of $\overline{I}$ over $\overline{B}$. Thus by Theorem 2.8, the completion of I is equal to $\overline{H} \otimes_{\overline{B}} \overline{I} = \overline{\overline{H}\overline{I}}$, and hence is a regular ideal of $\overline{H}$.

Let $\overline{M} = M/(R \cap \mathcal{D})$; then $\overline{M}$ is the maximal ideal of $\overline{R}$, and $\overline{\overline{B}\overline{M}} = \overline{BM}$ is the maximal ideal of the quasi-local ring $\overline{B}$. The maximal ideal of $\overline{H}$ is $\overline{\overline{H}\overline{M}}$, and since $\overline{\overline{H}\overline{I}}$ is a regular ideal of $\overline{H}$, it follows that $\overline{\overline{H}\overline{I}}$ is an $\overline{\overline{H}\overline{M}}$-primary ideal of $\overline{H}$. In particular, for any $n > 0$ we have $B/(BM)^n \cong \overline{B}/(\overline{B}\overline{M})^n \cong \overline{H}/(\overline{H}\overline{M})^n$. Thus for large n we have $\mathcal{L}(B/(BM)^n) = e_{\overline{H}}n - \rho_{\overline{H}}$. We therefore define the multiplicity and reduction number of B (even though B is not a Noetherian local ring) by:

$$e_B = e_{\overline{H}} \text{ and } \rho_B = \rho_{\overline{H}}.$$

We similarly define $e_{\overline{B}} = e_{\overline{H}}$ and $\rho_{\overline{B}} = \rho_{\overline{H}}$. Now $H \otimes_R B$ is a strongly unramified ring extension of H in $H \otimes_R Q$, and $(H \otimes_R B)/H \cong B/R$. Thus $J = \mathrm{Hom}_R(K,B/R) \cong \mathrm{Hom}_H(K,(H \otimes B)/H)$, and hence the completion of $H \otimes_R B$ is equal to $H/J = \overline{H}$. Therefore, we have $e_{H \otimes_R B} = e_{\overline{H}} = e_B$ and

$\rho_H \otimes_R B = \rho_{\overline{H}} = \rho_B$.

We shall say that an element $y \in B$ is superficial of degree s in B if $y \in (BM)^s$ and $(BM)^n y = (BM)^{n+s}$ for all large values of n. We can then define the first neighborhood ring Γ of B to be the set of all elements of Q of the form $\gamma = w/y$, where $w \in (BM)^s$ and y is a superficial element of B of degree s (s arbitrary).

With the remarks and definitions we have made it is not hard to show that Theorem 12.1 remains true if we replace R in that theorem by a strongly unramified ring extension B of R in Q ((4) of Theorem 12.1 is modified to read that B is equal to its first neighborhood ring if and only if B is a pseudo valuation ring).

Theorem 14.4. _Let B be a strongly unramified ring extension of R in Q, and let Λ be the first neighborhood ring of R. Then $B\Lambda = B + \Lambda$ is the first neighborhood ring of B._

Proof. By Theorem 12.1 there is a superficial element a in R of degree s such that $\Lambda = M^s a^{-1}$. Since $(BM)^n a = (BM)^{n+s}$ for all large n we see that a is a superficial element of B of degree s. We have $B\Lambda = (BM)^s a^{-1}$ and thus every element of $B\Lambda$ is of the form x/a, where $x \in (BM)^s$ and a is superficial in B of degree s. Therefore by definition, $B\Lambda \subset \Gamma$, the first neighborhood ring of B.

Because $(B\Lambda)a = (BM)^s$, we have $\Gamma a = \Gamma(B\Lambda)a = \Gamma(BM)^s$. By Theorem 12.1 applied to B and Γ, there is an integer $n > 0$ such that $\Gamma(BM)^{ns} = (BM)^{ns}$. Thus we have $\Gamma a^n = (\Gamma(BM)^s)^n = \Gamma(BM)^{ns} = (BM)^{ns} = (B\Lambda)a^n$. Cancelling a^n we obtain $\Gamma = B\Lambda$. By Theorem 6.1 we have $B\Lambda = B + \Lambda$.

Remarks. It is an immediate consequence of Theorem 14.4 that Theorem 12.2 is true for any strongly unramified ring extension B of R in Q. It is also easy to see that Theorem 12.3 is true for B. Because Theorems 12.4, 12.5 and 12.6 depend only on Theorems 12.1, 12.2 and 12.3, they are also true for B. Thus in proving the following theorems we can freely apply Theorems 12.1 through 12.6 to strongly unramified ring extensions of R in Q.

<u>Theorem 14.5</u>. <u>Let</u> A <u>be a closed component of</u> R <u>and let</u> B/R <u>be a simple divisible</u> R-<u>submodule of</u> K <u>corresponding to</u> A. <u>If</u> a <u>is a regular element of</u> R, <u>then</u>:

$$\mathcal{L}(R/Ra) = \mathcal{L}(B/Ba) + \mathcal{L}(A/Aa).$$

<u>Proof</u>. We have an exact sequence:

$$0 \to (Ba \cap R)/Ra \to R/Ra \to R/(Ba \cap R) \to 0.$$

and thus $\mathcal{L}(R/Ra) = \mathcal{L}(R/(Ba \cap R)) + \mathcal{L}((Ba \cap R)/Ra)$. By Theorem 6.1 we have $R/(Ba \cap R) \cong B/Ba$, and thus it is sufficient to prove that $\mathcal{L}((Ba \cap R)/Ra) = \mathcal{L}(A/Aa)$.

Now $(Ba \cap R)/Ra \cong (B \cap Ra^{-1})/R$, and we have an exact sequence:

$$0 \to (B \cap Ra^{-1})/R \to B/R \xrightarrow{f} B/R \to 0$$

where f is multiplication by a. By Theorem 8.2, B/R is isomorphic to Q/C, where C is a regular ideal of A. Thus from the preceding exact sequence, we see that $(B \cap Ra^{-1})/R$ is isomorphic to the kernel of $g : Q/C \to Q/C$, where g is multiplication by a. Since $\mathrm{Ker}\ g = Ca^{-1}/C$ and $Ca^{-1}/C \cong C/aC$, we see that $(Ba \cap R)/Ra \cong C/aC$. By Theorem 12.3 we have $\mathcal{L}(C/aC) = \mathcal{L}(A/aA)$, and this concludes the proof of the theorem.

<u>Corollary 14.6</u>. <u>Let</u> A <u>be a closed component of</u> R <u>and let</u> B/R <u>be a simple divisible</u> R-<u>submodule of</u> K <u>corresponding to</u> A. <u>Then</u>:

$$e = e_B + e_A.$$

<u>Proof</u>. Let a be a superficial element of R of degree s. As we have seen in the proof of Theorem 14.4, a is superficial of degree s in both B and A. By Theorem 12.5 we see that $\mathcal{L}(R/Ra) = se$,

$\mathcal{L}(B/Ba) = se_B$, and $\mathcal{L}(A/Aa) = se_A$. The corollary now follows from Theorem 14.5.

<u>Corollary 14.7</u>. <u>Let</u> $B \neq R$ <u>be a strongly unramified ring extension of</u> R <u>in</u> Q. <u>Then</u> $e_B < e$.

<u>Proof</u>. By the Remarks preceding Theorem 14.4, we can assume without loss of generality that R is a complete ring. Let B_1/R be a simple divisible R-submodule of B/R. By Corollary 14.6 we have $e_{B_1} < e$. If $\mathcal{D}_1$ is the divisible submodule of B_1, let $\overline{B}_1 = B_1/\mathcal{D}_1$ and $\overline{B} = B/\mathcal{D}_1$. Now $B_1 = R + \mathcal{D}_1$, and hence $\overline{B}_1 \cong R/(R \cap \mathcal{D}_1)$ is a complete, local, 1-dimensional Cohen-Macaulay ring. Since $\overline{B}$ is a strongly unramified ring extension of $\overline{B}_1$, we have by induction on L(K) that $e_{\overline{B}} \leq e_{\overline{B}_1}$. However, $e_B = e_{\overline{B}}$ and $e_{B_1} = e_{\overline{B}_1}$, and thus $e_B \leq e_{B_1} < e$.

<u>Definition</u>. Let $P_1, \ldots, P_n$ be the prime ideals of rank 0 in H and let $N_1, \ldots, N_n$ be the corresponding P_i-primary components of 0 in H. An ideal I of H will be called a <u>minimal ideal for</u> P_i if $I = J_i \cap (\bigcap_{j \neq i} N_j)$ where $J_i = H$ if $N_i = P_i$, or if $N_i \neq P_i$, then J_i is a P_i-primary ideal of H such that $J_i \supsetneq N_i$ and there are no P_i-primary ideals properly between J_i and N_i. If 0 is a prime ideal of H, then by convention H is the minimal ideal for 0.

<u>Theorem 14.8</u>. <u>Let</u> P_i <u>be a prime ideal of rank</u> 0 <u>in</u> H. <u>Then there is a one-to-one correspondence between the set of simple divisible submodules</u> D <u>of</u> K <u>corresponding to</u> P_i <u>and the set of minimal ideals</u> I <u>for</u> P_i <u>given by</u> $D = K \otimes_R I$ <u>and</u> $\mathrm{Hom}_R(K,D) = I$.

<u>Proof</u>. If $P_i = 0$, then H is an integral domain, and hence by Theorem 7.1, K is a simple divisible R-module. In this case the theorem reduces to the assertions that $K \otimes_R H = K$ and $\mathrm{Hom}_R(K,K) = H$. Thus we can assume that $P_i \neq 0$.

If I is a minimal ideal for P_i, then I is a minimal unmixed ideal of rank 0 in H. Thus by Theorem 6.6, $D = K \otimes_R I$ is a simple divisible submodule of K. Since $P_i I = 0$, we see that $P_i D = 0$, and hence $\mathrm{Ann}_H D = P_i$. Therefore, D corresponds to P_i. On the other hand if D is a simple divisible submodule of K corresponding to P_i, then by Theorem 6.6, $I = \mathrm{Hom}_R(K,D)$ is a minimal, nonzero, unmixed ideal of rank 0 in H. If J and L are unmixed ideals of rank 0 in H, then $J \subset L$ if and only if the P_j-primary component of J is contained in the P_j-primary component of L for $j = 1,\ldots,n$. Thus I is a minimal ideal for some P_k. Since $P_i D = 0$, we have $P_i I = 0$, and therefore $k = i$. Hence I is a minimal ideal for P_i.

The one-to-one nature of this correspondence follows from Theorem 6.6.

Remarks. If $\mathcal{S}$ is the set of zero divisors in H, then $H \otimes_R Q = H_{\mathcal{S}}$. We shall show that every simple divisible R-module is equivalent to an R-module of the form $I_{\mathcal{S}}/I$, where I is a minimal ideal for some rank 0 prime ideal of H. For let D be a simple divisible R-module, and let P be the prime ideal of rank 0 in H corresponding to D. By Corollary 8.9, D is equivalent to a simple divisible submodule B/R of K. By Theorem 14.8, $B/R = K \otimes_R I$, where I is a minimal ideal for P. Now $B/R \cong (H \otimes_R B)/H$ by Theorem 2.10. If $\mathcal{B}$ is the divisible submodule of $H \otimes_R B$, then by Theorem 6.8, $\mathcal{B}$ is an ideal of $H \otimes_R Q$; $H \otimes_R B = H + \mathcal{B}$; and $I = \mathcal{B} \cap H$. Thus we have $(H \otimes_R B)/H = (H + \mathcal{B})/H \cong \mathcal{B}/I$. Clearly $\mathcal{B} = (H \otimes_R Q)I = I_{\mathcal{S}}$, and hence D is equivalent to $I_{\mathcal{S}}/I$. We note finally that $I_{\mathcal{S}}/I$ is the torsion submodule of $(H \otimes_R Q)/I$.

The next theorem was proved originally by Northcott for the analytic components of R [21, Theorem 17].

Theorem 14.9. (Northcott). Let $A_1,\ldots,A_n$ be the closed components of R and let $P_1,\ldots,P_n$ be their corresponding prime ideals of rank 0 in H. Then:

$$e = \sum_{i=1}^{n} \mu(P_i)e_{A_i}.$$

Proof. It is clear from the remarks preceding Theorem 14.4 that we can assume that R is complete. If $L(K) = 1$, then R is a closed domain and the theorem is trivial. Hence we can assume that $L(K) > 1$. Let B/R be a simple divisible R-submodule of K corresponding to A_1. By Corollary 14.6 we have $e = e_B + e_{A_1}$. Hence it is sufficient to prove that $e_B = (\mu(P_1) - 1)e_{A_1} + \sum_{i>2} \mu(P_i)e_{A_i}$.

Let $\mathcal{D}$ be the divisible submodule of B; and $\overline{B} = B/\mathcal{D}$. By Theorem 6.8, $B = R + \mathcal{D}$, and hence $\overline{B} \cong R/(\mathcal{D} \cap R)$. Thus $\overline{B}$ is a complete, local, 1-dimensional Cohen-Macaulay ring. By Theorem 14.8, $B/R = K \otimes_R I$, where I is a minimal ideal for P_1. By Theorem 6.8, $\overline{B} = R/I$, and thus $I = \mathcal{D} \cap R$. It is obvious from the definition of a minimal ideal for P_1 that the latent multiplicities of $\overline{B}$ are $(\mu(P_1) - 1), \mu(P_2),\dots,\mu(P_n)$. Hence, for $i > 1$ we have $B \subset A_i$; and $B \subset A_1$ if and only if $(\mu(P_1) - 1) \neq 0$. If we let $\overline{A_i} = A_i/\mathcal{D}$, then $\overline{A_2},\dots,\overline{A_n}$ are closed components of $\overline{B}$ in $\overline{Q} = Q/\mathcal{D}$; and $\overline{A_1}$ is a closed component of $\overline{B}$ if and only if $(\mu(P_1) - 1) \neq 0$. Therefore, by induction on $L(K)$ we have $e_{\overline{B}} = (\mu(P_1) - 1)e_{\overline{A_1}} + \sum_{i>2} \mu(P_i)e_{\overline{A_i}}$. Since $e_B = e_{\overline{B}}$ and $e_{A_i} = e_{\overline{A_i}}$, the theorem has been proved.

Corollary 14.10. Let $V_1,\dots,V_n$ be the pseudo valuation rings of R and $P_1,\dots,P_n$ the corresponding prime ideals of rank 0 in H. Then:

$$e = \sum_{i=1}^{n} \mu(P_i)\mathcal{L}(V_i/MV_i)$$

Proof. Let $A_1,\dots,A_n$ be the closed components of R corresponding to $P_1,\dots,P_n$. By Theorem 14.9 it is sufficient to prove that $e_{A_i} = \mathcal{L}(V_i/MV_i)$. Let $\mathcal{D}_i$ be the divisible submodule of A_i and let $\overline{A_i} = A_i/\mathcal{D}$ and $\overline{V_i} = V_i/\mathcal{D}$. By Theorem 7.2, $\overline{A_i}$ is an analytically irreducible, Noetherian, local domain of Krull dimension 1; and by Corollary 7.6, $\overline{V_i}$ is the integral closure of $\overline{A_i}$ and $\overline{V_i}$ is a discrete valuation ring. Since $e_{A_i} = e_{\overline{A}_i}$ and $\overline{V}_i/(\overline{A_i M}\ \overline{V}_i) \cong V_i/MV_i$, we can assume without loss of generality that R is an analytically irreducible, Noetherian local domain of Krull dimension 1. If V is the integral closure of R, then V is a discrete valuation ring that is a finitely generated R-module.

If Λ is the first neighborhood ring of R, then $\Lambda \subset V$, since Λ is integral over R. Therefore, Λ is a local ring. Hence by Theorem 12.9, R has a superficial element a of degree 1. By Theorem 12.1 we have $\Lambda a = \Lambda M$, and hence $Va = V\Lambda a = V\Lambda M = VM$. Thus by Theorem 12.3 we have $\mathcal{L}(V/VM) = \mathcal{L}(\Lambda/\Lambda M)$. Since $\mathcal{L}(\Lambda/\Lambda M) = e$ by Theorem 12.4, the corollary is proved.

Definition. With the notation of Corollary 14.10, $V_i/\mathcal{D}_i$ is a discrete valuation ring. If $M_i/\mathcal{D}_i$ is the maximal ideal of $V_i/\mathcal{D}_i$, then M_i is a regular maximal ideal of V_i and contains every regular ideal of V_i. We define $f_i = [V_i/M_i : R/M]$, the dimension of the field extension V_i/M_i over the residue field R/M. Then the f_i's are called the latent residue degrees of R. Let ℓ_i be the length of V_i/MV_i as a V_i-module, so that $f_i \ell_i = \mathcal{L}(V_i/MV_i)$; and let $\mu_i = \mu(P_i)$.

Corollary 14.11. $e = \sum_{i=1}^{n} \mu_i f_i \ell_i$.

Proof. This is an immediate consequence of Corollary 14.10.

Definition. Let V_i be a pseudo valuation ring of R and let $\mathcal{D}_i$ be the divisible submodule of V_i. Let v_i be the valuation associated with the discrete valuation ring $V_i/\mathcal{D}_i$. If y is a regular element of V_i, let $\overline{y} = y + \mathcal{D}_i$ and define $v_i(y) = v_i(\overline{y})$.

The next theorem is due to Northcott [19, Th. 6].

Theorem 14.12. If b is a regular element of R, then

$$\mathcal{L}(R/Rb) = \sum_{i=1}^{n} \mu_i f_i v_i(b).$$

Proof. Using Theorem 14.5 we prove that $\mathcal{L}(R/Rb) = \sum_{i=1}^{n} \mu_i \mathcal{L}(A_i/A_i b)$ in the same way that we proved Theorem 14.9. By Theorem 12.3 applied to A_i we have $\mathcal{L}(A_i/A_i b) = \mathcal{L}(V_i/V_i b)$. Since $\mathcal{P}_i \subset V_i b$, we see that $V_i/V_i b \cong \overline{V}_i/\overline{V}_i\overline{b}$. Thus it is clear that $\mathcal{L}(V_i/V_i b) = f_i v_i(b)$.

Theorem 14.13. We have $\rho \geq L(K) - 1$; and if R has a closed component that is not a pseudo valuation ring, then $\rho \geq L(K)$.

Proof. By Theorem 12.14, we have $\rho \geq e - 1$; by Theorem 14.2 we have $L(K) = \sum_{i=1}^{n} \mu(P_i)$; and by Theorem 14.9, we have $e = \sum_{i=1}^{n} \mu(P_i) e_{A_i}$. Now $e_{A_i} = e_{\overline{A}_i}$, and $\overline{A}_i$ is a local, 1-dimensional, Cohen-Macaulay ring. Hence by Theorem 12.16, $e_{A_i} = 1$ if and only if $\overline{A}_i$ is a discrete valuation ring; that is, if and only if A_i is a pseudo valuation ring. The proof of the theorem is now obvious.

Theorem 14.14. If $B \neq R$ is a strongly unramified ring extension of R in Q, then $\rho_B < \rho$.

Proof. By Theorem 6.1, B/R is a proper, nonzero, divisible R-submodule of K, and thus $L(K) > 1$. In particular, R is not a discrete valuation ring. By Theorem 12.4 we have $\rho = \mathcal{L}(\Lambda/R)$, where Λ is the first neighborhood ring of R. By Theorem 14.4 the first neighborhood ring of B is $B\Lambda$, and thus we have $\rho_B = \mathcal{L}(B\Lambda/B)$ by Theorem 12.4. Since $B\Lambda = B + \Lambda$, we see that $B\Lambda/B \cong \Lambda/(\Lambda \cap B)$. Thus we need to prove that $\mathcal{L}(\Lambda/(\Lambda \cap B)) < \mathcal{L}(\Lambda/R)$. Hence it is sufficient to prove that $(\Lambda \cap B)/R \neq 0$.

Since R is not a discrete valuation ring, we have $M^{-1} \subset \Lambda$ by Theorem 12.13. Now $\mathrm{Ann}_{B/R} M \subset M^{-1}/R$; and $\mathrm{Ann}_{B/R} M \neq 0$, since B/R is an Artinian R-module. Thus $\mathrm{Ann}_{B/R} M \subset (M^{-1} \cap B)/R \subset (\Lambda \cap B)/R$, and hence $(\Lambda \cap B)/R \neq 0$.

Theorem 14.15: Assume that the maximal ideal of R can be generated by two elements. Let A be a closed component of R and B/R a *simple divisible R-submodule of K corresponding to A. Then:*

$$\rho = \rho_B + \rho_A + e_B e_A.$$

Proof. If $B = Q$, then $\rho_B = 0 = e_B$ and $A = R$. The equation is trivial in this case. Thus we may assume that $B \neq Q$.

Since the completions $H(B)$ and $H(A)$ of B and A, respectively, are homomorphic images of H by Theorem 6.6, they are local, 1-dimensional, Cohen-Macaulay rings whose maximal ideals can be generated by two elements. By the Remarks preceding Theorem 14.4, the multiplicities and reduction numbers of B and A are unchanged upon passage to $H(B)$ and $H(A)$. Thus by Corollary 13.8 we have $\rho = \frac{1}{2} e(e-1)$, $\rho_B = \frac{1}{2} e_B(e_B-1)$, and $\rho_A = \frac{1}{2} e_A(e_A-1)$. By Corollary 14.6 we have $e = e_B + e_A$. Therefore we have $\rho = \frac{1}{2} (e_B + e_A)(e_B + e_A - 1)$ $= \frac{1}{2} e_B(e_B-1) + \frac{1}{2} e_A(e_A-1) + e_B e_A = \rho_B + \rho_A + e_B e_A$.

Corollary 14.16. Let $A_1,\dots,A_n$ be the closed components of R and $\mu_1,\dots,\mu_n$ their associated latent multiplicities. Assume that the maximal ideal of R can be generated by two elements. Then:

$$\rho = \sum_i \mu_i \rho_{A_i} + \frac{1}{2} \sum_i \mu_i(\mu_i-1) e_{A_i} + \sum_{i \neq j} \mu_i \mu_j e_{A_i} e_{A_j}.$$

Proof. As we observed in the proof of Theorem 14.15 we have the following equations by Corollary 13.8:

$$\rho = \frac{1}{2} e(e-1) \text{ and } \rho_{A_i} = \frac{1}{2} e_{A_i}(e_{A_i}-1) \text{ for } i = 1,\dots,n.$$

Thus by Theorem 14.9 we have $\rho - \frac{1}{2} e^2 = -\frac{1}{2} e = -\frac{1}{2} \sum_i \mu_i e_{A_i}$ $= \sum_i \mu_i(\rho_{A_i} - \frac{1}{2} e_{A_i}^2)$. Therefore,

$$\begin{aligned} \rho &= \frac{1}{2} e^2 + \sum_i \mu_i \rho_{A_i} - \frac{1}{2} \sum_i \mu_i e_{A_i}^2 \\ &= \frac{1}{2}(\sum_i \mu_i e_{A_i})^2 - \frac{1}{2} \sum_i \mu_i e_{A_i}^2 + \sum_I \mu_i \rho_{A_i} \\ &= \sum_i \mu_i \rho_{A_i} + \sum_{i \neq j} \mu_i \mu_j e_{A_i} e_{A_j} + \frac{1}{2} \sum_i (\mu_i^2 - \mu_i) e_{A_i}^2. \end{aligned}$$

Definition. We shall say that a Noetherian local domain is analytically ramified if it is not analytically unramified. By Theorem 10.2, a Noetherian local domain of Krull dimension 1 is analytically ramified if and only if its integral closure is not a finitely generated module. Nagata [17] has given several examples of analytically ramified Noetherian local domains of Krull dimension 1. Hence the following theorem proves the existence of analytically ramified Gorenstein local domains of Krull dimension 1.

Theorem 14.16. *If R is an analytically ramified Noetherian local domain of Krull dimension 1, then there exists an analytically ramified local Gorenstein ring T of Krull dimension 1 such that* $R \subset T \subset Q$.

Proof. Let $0 = N_1 \cap \ldots \cap N_n$ be a normal decomposition of 0 in H where N_i is a P_i-primary ideal of H and $P_1, \ldots, P_n$ are the prime ideals of rank 0 in H. Let V_i be the valuation ring of R corresponding to P_i, and let $K \otimes_R N_i = B_i/R \subset K$. By Theorem 11.8, B_i is a strongly unramified extension of R in Q, $B_i \subset V_i$, and $B_i \not\subset V_j$ for $i \neq j$. Thus V_i is the integral closure of B_i.

If S is the integral closure of R, then $L(S/R) > 0$ by Theorem 5.1. Hence by Corollary 14.3, there is an index i such that $u(P_i) > 1$. It is clear from the proof of Theorem 14.2 that $L(Q/B_i) = u(P_i)$. Thus $L(V_i/B_i) = L(Q/B_i) - L(Q/V_i) = u(P_i) - 1 > 0$. Hence B_i is analytically ramified. Therefore, without loss of generality, we may assume that $R = B_i$. Hence the integral closure of R is a valuation ring V and $L(V/R) \geq 1$.

Now there exists a divisible submodule B/R of K such that $L(Q/B) = 2$. By Theorem 6.1, B is a strongly unramified extension ring of R. Since the integral closure of B is V, B has only one closed component A. By Theorems 7.5 and 5.1, V is a finitely generated A-module. If we adjoin these generators to B we obtain an analytically ramified local domain T (of Krull dimension 1) such that $L(Q/T) = 2$ and V is the only closed component of T.

The completion of T has only one rank 0 prime ideal P by Theorem 7.3. If $\mu(P)$ is the latent multiplicity of P, then by Theorem 14.2 we have $\mu(P) = L(Q/T) = 2$. Since $e_V = 1$, we see by Theorem 14.9 that $e_T = \mu(P) = 2$. Hence by Theorem 12.17, every ideal of T can be generated by 2 elements. Therefore, by Theorem 13.2, T is a Gorenstein ring.

Definition: If S is an Artinian local ring, then we shall denote the length of S as an S-module by $\lambda(S)$. If R is a 1-dimensional, local, Cohen-Macaulay ring and p is a prime ideal of rank 0 in R, we shall let $\mu(p)$ denote the length of a composition series of p-primary ideals between p and the p-primary component of 0 in R. It is then easy to see that $\mu(p) = \lambda(R_p)$.

Lemma 14.17: *Let* P *be a prime ideal of rank* 0 *in* H, $p = P \cap R$, *and* $I = Hp$. *Then* H/I *is the completion of* R/p, *and* $\mu(P) = \mu(p)\mu(P/I)$.

Proof. It is easily seen that H/I is the completion of R/p. Since $(H/I)_{P/I} \cong H_P/I_P$, what we have to prove is that $\lambda(H_P) = \lambda(R_p)\lambda(H_P/I_P)$. Let us take a composition series for R_p:

$$R_p = J_0 \supset J_1 \supset \ldots \supset J_n = 0$$

where the J_i's are ideals of R_p and $J_i/J_{i+1} \cong k$, the quotient field of R/p. Then we have a chain

$$H_P \otimes_R R_p = H_P \otimes_R J_0 \supset H_P \otimes_R J_1 \supset \ldots \supset H_P \otimes_R J_n = 0.$$

Since $H_P \otimes_R R_p \cong H_P$, it is sufficient to prove that $H_P \otimes_R k \cong H_P/I_P$. Let $\bar{R} = R/p$; then

$$H \otimes_R k \cong H \otimes_R (\bar{R} \otimes_{\bar{R}} k) \cong (H \otimes_R \bar{R}) \otimes_{\bar{R}} k \cong (H/I) \otimes_{\bar{R}} k.$$

Hence $H \otimes_R k$ is the full ring of quotients of H/I by Theorem 3.6. But if $P = P_1, \ldots, P_m$ are the rank 0 prime ideals of H that contain I, then $\sum_{j=1}^{m} \oplus H_{P_i}/I_{P_i}$ is also the full ring of quotients of H/I, and hence $H_P \otimes_R k \cong H_P \otimes_H (H \otimes_R k) \cong H_P \otimes_H (\sum_{i=1}^{m} \oplus H_{P_i}/I_{P_i}) \cong H_P/I_P$.

Theorem 14.18: <u>Let</u> $p_1, \ldots, p_t$ <u>be the prime ideals of rank</u> 0 <u>of</u> R <u>and let</u> $\bar{R}_j = R/p_j$. Then $e = \sum_{j=1}^{t} \mu(p_j) e_{\bar{R}_j}$.

<u>Proof</u>. Let $P_{j_1}, \ldots, P_{j_k}$ be the rank 0 prime ideals of H having the property that $P_{j_i} \cap R = p_j$; and let $A_{j_1}, \ldots, A_{j_k}$ be the closed components of R that correspond to the P_{j_i}'s. Now $Qp_j \subset A_{ji}$ for all $i = 1, \ldots, k$ and we let $\bar{A}_{j_i} = A_{j_i}/Qp_j$. Then the $\bar{A}_{j_i}$'s are the closed components of $\bar{R}_j$, and hence by Theorem 14.9

$e_{\bar{R}_j} = \sum_{i=1}^{k} \mu(P_{j_i}/Hp_j) e_{\bar{A}_{j_i}}$. Since $e_{\bar{A}_i} = e_{A_i}$, it follows from Lemma 14.17 and Theorem 14.9 that $e = \sum_{i,j} \mu(P_{j_i}) e_{A_{j_i}} = \sum_{i,j} \mu(p_j)\mu(P_{j_i}/Hp_j) e_{\bar{A}_{j_i}}$ $= \sum_{j=1}^{t} \mu(p_j) e_{\bar{R}_j}$.

CHAPTER XV

THE CANONICAL IDEAL OF R

Throughout this chapter R will be a local, 1-dimensional, Cohen-Macaulay ring. We shall explore further than we have already done the relationship between the modules K and E. We shall then use these relationships to establish necessary and sufficient conditions for the existence of a canonical ideal of R.

Theorem 15.1. *There is a one-to-one, order-reversing correspondence between the set of nonzero divisible submodules D of E and the set of unmixed ideals I of rank 0 in H given by:*

$$\mathrm{Ann}_H D = I \text{ and } \mathrm{Ann}_E I = D.$$

Proof. Let D be a nonzero divisible submodule of E and let $I = \mathrm{Ann}_H E$. Suppose that h is an element of H and b a regular element of R such that $bh \in I$. Then $0 = hbD = hD$, and hence $h \in I$. This proves that H/I is a torsion-free R-module. Therefore, by Lemma 6.5, I is an unmixed ideal of rank 0 in H.

Conversely, let I be an unmixed ideal of rank 0 in H and let $D = \mathrm{Ann}_E I$. Then $D \cong \mathrm{Hom}_H(H/I,E)$, and hence D is the injective envelope over H/I of the residue field of H/I. Thus D is divisible by the regular elements of H/I. If b is a regular element of R, then b is a regular element of H, and we let $\bar{b} = b + I$, an element of H/I. Since Hb is an HM-primary ideal of H and I is an unmixed ideal of rank 0 in H, it follows that $\bar{b}$ is a regular element of H/I. Thus D is divisible by $\bar{b}$, and hence by b. Therefore, D is a nonzero, divisible R-module.

The correspondence we have established is a one-to-one, order-reversing correspondence because of the duality between the ideals of H and the submodules of E that is given in Theorem 4.5.

Corollary 15.2. *There is a one-to-one, order-reversing corres-*

pondence between the set of divisible submodules of K *and the set of divisible submodules of* E. *Thus* $L(K) = L(E)$.

Proof. By Theorem 6.6 there is a one-to-one, order-preserving correspondence between the set of proper divisible submodules of K and the set of unmixed ideals of rank 0 in H. This fact combined with Theorem 15.1 gives an immediate proof of the corollary.

Corollary 15.3. *There is a one-to-one, order-reversing correspondence between the set of nonzero, divisible submodules* D *of* E *and the set of strongly unramified extensions* B *of* R *in* Q *such that if* D *and* B *correspond, then* D *is isomorphic to the torsion submodule of* $E_B(B/MB)$, *where* $E_B(B/MB)$ *is the injective envelope over* B *of* B/MB. *If* R *has no nilpotent elements other than* 0, *then* $D \cong E_B(B/MB)$.

Proof. Let D be a nonzero divisible submodule of E, and let $I = \mathrm{Ann}_H D$. By Theorem 15.1, I is an unmixed ideal of rank 0 in H. If $B/R = K \otimes_R I$, then B is a strongly unramified ring extension of R in Q, and H/I is the completion of B by Theorem 6.6. The residue field of H/I is isomorphic to B/MB; and in the proof of Theorem 15.1 we have seen that D is the injective envelope over H/I of B/MB. Thus D is an essential extension over B of B/MB and hence we have an imbedding $D \subset E_B(B/MB)$. Because D is a torsion B-module, we have $D \subset t(E_B(B/MB))$. However, $t(E_B(B/MB))$ is an H/I-module by Theorem 2.7, and thus D is a direct summand of $t(E_B(B/MB))$ since D is an injective H/I-module. Since $E_B(B/MB)$ is an essential extension of B/MB, it follows that $D = t(E_B(B/MB))$.

This correspondence between D and B that we have found is clearly a one-to-one order-reversing correspondence by Theorems 6.6 and 15.1. If R has no nilpotent elements, other than 0, then as we have seen in Corollary 4.4, Q is a semi-simple ring. Because Q is also the full ring of quotients of B, it follows from Theorem 1.4 that $t(E_B(B/MB))$ is a direct summand of $E_B(B/MB)$. Since $E_B(B/MB)$ is an essential extension of B/MB this implies that $E_B(B/MB)$ $= t(E_B(B/MB))$. Therefore, by the preceding paragraph we have $D \cong E_B(B/MB)$.

Theorem 15.4. K *and* E *have equivalent composition series of divisible* R-*modules*.

Proof. Let P be a prime ideal of rank 0 in H, and let E_P and K_P be the P-primary components of E and K, respectively. By Theorem 11.10, it is sufficient to prove that $L(K_P) = L(E_P)$. If μ is the latent multiplicity of P, then by Theorem 14.2, we have $L(K_P) = \mu$. Hence it is sufficient to prove that $L(E_P) = \mu$ also.

Let N be the P-primary component of 0 in H. Then we have a composition series of P-primary ideals:

$$P = N_1 \supsetneqq N_2 \supsetneqq \cdots \supsetneqq N_\mu = N.$$

Putting $D_i = \mathrm{Ann}_E N_i$, we have by Theorem 15.1 a chain of divisible submodules of E:

$$0 \neq D_1 \subsetneqq D_1 \subsetneqq D_2 \subsetneqq \cdots \subsetneqq D_\mu.$$

By Theorem 4.5, $\mathrm{Ann}_H D_i = N_i$, and thus every D_i is a P-primary divisible R-module. Hence we have $D_\mu \subset E_P$.

On the other hand, $\mathrm{Ann}_H E_P$ is a P-primary ideal of H by definition, and thus $N \subset \mathrm{Ann}_H E_P$. It follows from Theorem 15.1 that $E_P \subset D_\mu$, and hence $D_\mu = E_P$. Since it is an immediate consequence of Theorem 15.1 that $L(D_\mu) = \mu$, we see that $L(E_P) = \mu = L(K_P)$, and the theorem is proved.

Definition. Let I be a fixed regular ideal of R, and let A be an R-module. Then we shall denote the R-module $\mathrm{Hom}_R(A,I)$ by A^*. There is a canonical R-homomorphism $\varphi : A \to A^{**}$ given by $\varphi(x)(f)$ $= f(x)$ for all $x \in A^*$.

If A is an R-submodule of Q, let us define

$A^I = \{q \in Q \mid qA \subset I\}$. If A contains a regular element of R, we shall show that there is a natural isomorphism of A* and A^I. For we have an exact sequence:

$$(\#) \qquad \mathrm{Hom}_R(Q/A,Q) \to \mathrm{Hom}_R(Q,Q) \to \mathrm{Hom}_R(A,Q) \to \mathrm{Ext}^1_R(Q/A,Q).$$

Since Q is a flat R-module, we have by [3, Ch. V, Prop. 4.1.3] that $\mathrm{Hom}_R(Q/A,Q) \cong \mathrm{Hom}_Q(Q \otimes_R Q/A,Q)$ and $\mathrm{Ext}^1_R(Q/A,Q) \cong \mathrm{Ext}^1_Q(Q \otimes_R Q/A,Q)$. But Q/A is a torsion R-module because A contains a regular element of R, and hence $Q \otimes_R Q/A = 0$ by Theorem 1.1. Thus we have $\mathrm{Hom}_R(Q/A,Q) = 0 = \mathrm{Ext}^1_R(Q/A,Q)$. It follows from exact sequence (#) that $\mathrm{Hom}_R(Q,Q) \cong \mathrm{Hom}_R(A,Q)$. Therefore, every R-homomorphism from A into Q is achieved by the multiplication on A by a unique element of Q. It follows from this that there is a natural isomorphism between A* and A^I.

If J is a regular ideal of R, then clearly J^I contains a regular element of R and is isomorphic to a regular ideal of R. Thus $(J^I)^*$ is naturally isomorphic to $(J^I)^I$, and we can identify the mapping $\varphi : J \to J^{**}$ with the containment $J \subset (J^I)^I$. If L is an ideal of R such that $J \subset L$, then obviously we have $L^I \subset J^I$.

The ideal I is said to be a <u>canonical ideal for R</u> if $\varphi : J \to J^{**}$ is an isomorphism for every regular ideal J of R; that is, if $J = (J^I)^I$. In particular, we then have $\mathrm{Hom}_R(I,I) = (R^I)^* \cong (R^I)^I = R$. It has been shown in Theorem 13.1 that R is a Gorenstein ring if and only if R is a canonical ideal for R. The next theorem is a generalization of this result.

<u>Theorem 15.5. Let</u> I <u>be a regular ideal of</u> R. <u>Then the following statements are equivalent:</u>

(1) I <u>is a canonical ideal for</u> R.

(2) $M^I/I \cong R/M$.

(3) $\mathrm{Ext}^1_R(R/M,I) \cong R/M$.

(4) $Q/I \cong E$

(5) Q/I _is an injective_ R-_module_.

(6) Q _and_ Q/I _are injective_ R-_modules_.

(7) $\text{inj.dim.}_R I = 1$.

Proof. (1) $\Longrightarrow$ (2). Let L be an R-module such that $I \subset L \subset M^I$. Since $I^I = (R^I)^I = R$ and $(M^I)^I = M$, we see that $M \subset L^I \subset R$. Thus $L^I = M$, or $L^I = R$; and hence $L = (L^I)^I$ is equal to either M^I or I. Therefore, M^I/I is a simple R-module, and hence $M^I/I \cong R/M$.

(2) $\Longleftrightarrow$ (3). We have an exact sequence:

$$\text{Hom}_R(R/M,Q) \to \text{Hom}_R(R/M,Q/I) \to \text{Ext}_R^1(R/M,I) \to \text{Ext}_R^1(R/M,Q).$$

The end terms of this sequence are annihilated by M, and yet are torsion-free and divisible. Thus the end terms of this sequence are 0, showing that we have an isomorphism: $\text{Ext}_R^1(R/M,I) \cong \text{Hom}_R(R/M,Q/I)$.

The socle of an R-module is defined to be the sum of all of its simple submodules. It is clear that the socle of Q/I is equal to M^I/I, and is isomorphic to $\text{Hom}_R(R/M,Q/I)$. Thus we have an isomorphism: $\text{Ext}_R^1(R/M,I) \cong M^I/I$. The equivalence of (2) and (3) is now apparent.

(2) $\Longrightarrow$ (4). Because I is a regular ideal, R/I is an Artinian R-module. Hence from the exact sequence:

$$0 \to R/I \to Q/I \to K \to 0$$

we see that Q/I is an Artinian R-module. Every Artinian R-module is an essential extension of its socle. Clearly the socle of Q/I is M^I/I, and hence by (2) is isomorphic to R/M. Every essential extension of R/M can be imbedded in E, and thus we can assume that $Q/I \subset E$. But from the preceding exact sequence we see that $L(Q/I) = L(K)$; and $L(K) = L(E)$ by Corollary 15.2. Therefore, $L(Q/I) = L(E)$, and hence $Q/I \cong E$.

(4) $\Longrightarrow$ (2). Since the socle of Q/I is equal to M^I/I, and the

socle of E is equal to R/M, it follows that if $Q/I \cong E$, then $M^I/I \cong R/M$.

(4) $\Longrightarrow$ (1). We shall show first that $(R^I)^I = R$. Let $q \in (R^I)^I$; then $qI \subset I$. Hence we can define an R-homomorphism $f : Q/I \to Q/I$ by $f(x + I) = qx + I$ for all $x \in Q$. Now $\mathrm{Hom}_R(E,E) = H$ by Theorem 4.5; and every R-submodule of E is an H-module by Theorem 2.7. Thus because $f \in H$, we have $f(R/I) \subset R/I$. It follows that there exists an element $r \in R$ such that $q + I = f(1+I) = r + I$. Therefore, $q \in R$, showing that $(R^I)^I = R$.

If b is any regular element of R, then $Rb = ((Rb)^I)^I$. For suppose that $q \in ((Rb)^I)^I$. Since $(Rb)^I = b^{-1}I$, we have $qb^{-1}I \subset I$. Thus $qb^{-1} \in I^I = (R^I)^I = R$, and hence $q \in Rb$.

Suppose that J is a regular ideal of R such that $J = (J^I)^I$, and let L be an ideal of R such that $L \supset J$ and $L/J \cong R/M$. We shall show that $L = (L^I)^I$. We have an exact sequence:

$$\mathrm{Hom}_R(R/M,I) \to \mathrm{Hom}_R(L,I) \to \mathrm{Hom}_R(J,I) \to \mathrm{Ext}^1_R(R/M,I).$$

The first term of this sequence is 0 because I is torsion-free; and the last term is isomorphic to R/M because of the proven equivalence of (4) and (3). Since $\mathrm{Hom}_R(L,I) \cong L^I$ and $\mathrm{Hom}_R(J,I) \cong J^I$, this sequence becomes:

$$0 \to L^I \to J^I \xrightarrow{\alpha} R/M.$$

If $\alpha = 0$, then $L^I = J^I$, and hence $J = (J^I)^I = (L^I)^I \supset L$. This contradiction shows that $\alpha \neq 0$, and thus α is onto. A repetition of the preceding type of argument shows that we can derive from this sequence another:

$$0 \to (J^I)^I \to (L^I)^I \xrightarrow{\beta} R/M.$$

If $\beta = 0$, then $J = (J^I)^I = (L^I)^I \supset L$. This contradiction shows that β is onto. Now $L/J \subset (L^I)^I/J$, and the last sequence shows that

$(L^I)^I/J$ is a simple R-module. Therefore, we have $L = (L^I)^I$.

Now let J be any regular ideal of R, and let b be a regular element of J. Then we have a composition series:

$$J_0 = Rb \subset J_1 \subset \ldots \subset J_n = J$$

where $J_{i+1}/J_i \cong R/M$ for all $i < n$. It follows from the preceding paragraphs that we can climb up the ladder and obtain $(J^I)^I = J$. Therefore, I is a canonical ideal of R.

(4) <==> (5). It is trivial that (4) ==> (5), and hence assume that Q/I is an injective R-module. Since Q/I is an Artinian R-module, it follows from Theorem 4.6, that Q/I is isomorphic to a direct sum of copies of E. Since $L(Q/I) = L(K) = L(E)$, we see that this direct sum can have only one term, and thus $Q/I \cong E$.

(4) ==> (7). Since Q/I is a torsion R-module, we have by Theorem 1.1 that $E \cong Q/I \cong \mathrm{Tor}_1^R(K,Q/I) \cong K \otimes_R I$. Thus $K \otimes_R \mathrm{Hom}_R(K,\mathrm{Hom}_R(I,E)) \cong K \otimes_R \mathrm{Hom}_R(K \otimes_R I,E) \cong K \otimes_R \mathrm{Hom}_R(E,E)$ $\cong K \otimes_R H \cong K$. Now $\mathrm{Hom}_R(I,E)$ is a homomorphic image of $\mathrm{Hom}_R(R,E)$, and thus is a torsion divisible R-module. Thus by Corollary 1.2, the preceding isomorphisms show that $\mathrm{Hom}_R(I,E) \cong K$.

Let J be an ideal of R. Since $\mathrm{Hom}_R(I,E) \cong K$, we have by [3, Ch. VI, Prop. 5.3]:

$$\mathrm{Tor}_n^R(K,R/J) \cong \mathrm{Tor}_1^R(\mathrm{Hom}_R(I,E),R/J) \cong \mathrm{Hom}_R(\mathrm{Ext}_R^n(R/J,I),E)$$

for all $n \geq 0$. Since $\mathrm{Tor}_n^R(K,R/J) = 0$ for all $n \geq 2$, it follows that $\mathrm{Ext}_R^n(R/J,I) = 0$ for all $n \geq 2$, and hence $\mathrm{inj.dim}_R I = 1$.

(7) ==> (6). It is clearly sufficient to prove that Q is an injective R-module. As we observed in the proof of Theorem 13.1, if we apply the exact functor $Q \otimes_R \cdot$ to an injective resolution over R of I, we obtain an injective resolution over Q of $Q \otimes_R I$. But I is a regular ideal, and hence $Q \otimes_R I \cong Q$. Thus $\mathrm{inj.dim}_Q Q \leq 1$. As in the proof of Theorem 13.1, this implies that Q is an injective R-module.

(6) ⟹ (5). This is a trivial assertion.

This concludes the proof of the theorem.

Definition. A commutative Noetherian ring of Krull dimension 0 is called a Gorenstein ring of dimension 0 if it is self injective. In the proof of Theorem 13.1 we showed that Q is self-injective if and only if it is an injective R-module. Thus we have the following corollary.

Corollary 15.6. *If R has a canonical ideal, then Q is a Gorenstein ring of dimension* 0.

Remarks. Q being a Gorenstein ring of dimension 0 is a necessary condition for R to have a canonical ideal, but it is not a sufficient condition. For there is an example of a Noetherian local domain of Krull dimension 1 that does not have a canonical ideal (see [5]).

On the other hand, we shall see in the next theorem that $H \otimes_R Q$ being a Gorenstein ring of dimension 0 is a necessary and sufficient condition for R to have a canonical ideal. The next theorem also answers several other questions raised earlier in these notes.

Theorem 15.7. *The following statements are equivalent:*

(1) *R has a canonical ideal.*

(2) $H \otimes_R Q$, *the full ring of quotients of* H, *is a Gorenstein ring of dimension* 0.

(3) *If* P *is a prime ideal of rank* 0 *in* H, *then the* P-*primary component of* 0 *in* H *is an irreducible ideal.*

(4) *There is a regular ideal* I *of* Q *such that* $E \cong Q/I$.

(5) K *is equivalent to* E.

(6) *There is an* R-*homomorphism of* K *onto* E.

(7) K *contains only one simple divisible* R-*module from each equivalence class of such modules.*

<u>Proof</u>. (1) <==> (4). This has been proved in Theorem 15.5.

(4) <==> (6). Suppose that I is a regular ideal of R such that $E \cong Q/I$. If b is a regular element of I, then K is isomorphic to Q/Rb. Since $Rb \subset I$ we have an R-homomorphism of Q/Rb onto Q/I, and hence an R-homomorphism of K onto E. Conversely, suppose that f is an R-homomorphism of K onto E. If $C/R = \operatorname{Ker} f$, then $E \cong K/\operatorname{Ker} f \cong Q/C$. We have $L(C/R) = L(K) - L(E)$, and hence $L(C/R) = 0$ by Corollary 15.2. Thus C is a finitely generated R-module by Theorem 5.1. It follows that there exists a regular element $b \in R$ such that $bC = I$ is a regular ideal of R. We have $E \cong Q/C \cong Q/bC = Q/I$.

(6) <==> (5). Let f be an R-homomorphism of K onto E. As in the proof of (4) <==> (6) we see that Ker f is reduced. Hence K is equivalent to E by Lemma 5.8. That (5) implies (6) follows immediately from the definition.

(5) <==> (7). The assertion that (5) implies (7) is a consequence of Theorem 9.5. Conversely, suppose that K contains only one simple divisible R-module from each equivalence class of such modules. Let P be a prime ideal of rank 0 in H, and let D be the unique simple divisible submodule of K_P, the P-primary component of K. There is a nonzero R-homomorphism g of D into E. Because E is injective, g can be extended to an R-homomorphism f of K_P into E. By Corollary 11.3, $\operatorname{Im} f \subset E_P$, the P-primary component of E. Now Ker f is reduced, for otherwise it would contain D. Thus we have $L(K_P) = L(\operatorname{Im} f)$. By Theorem 15.4, K and E have equivalent composition series of divisible submodules, and thus $L(K_P) = L(E_P)$. Hence $L(\operatorname{Im} f) = L(E_P)$, and so $E_P = \operatorname{Im} f$. Therefore, K_P and E_P are equivalent by Corollary 5.13. It follows that if $P_1,\ldots,P_n$ are the prime ideals of rank 0 in H, then $K_{P_1} \oplus \ldots \oplus K_{P_n}$ is equivalent to $E_{P_1} \oplus \ldots \oplus E_{P_n}$. Since K and E are equivalent to these two direct sums, respectively, by Theorem 11.5, it follows that K and E are equivalent to each other.

(7) <==> (3). Let $P_1,\ldots,P_n$ be the prime ideals of rank 0 in

H, and let $N_1,\ldots,N_n$ be the P_i-primary components of 0 in H. Let D and D' be simple divisible submodules of K corresponding to P_1. Then by Theorem 14.8 we have $D = K \otimes_R I$ and $D' = K \otimes_R I'$, where I and I' are minimal ideals of H for P_1. By definition $I = J_1 \cap N_2 \cap \ldots \cap N_n$ and $I' = J_1' \cap N_2 \cap \ldots \cap N_n$, where $J_1 = H = J_1'$ if $P_1 = N_1$, and otherwise J_1 and J_1' are P_1-primary ideals of H that properly contain N_1 and are minimal with respect to that property. Now $h(D \cap D')$ $= h((K \otimes_R I) \cap (K \otimes_R I')) = K \otimes_R (I \cap I')$ $= K \otimes_R (J_1 \cap J' \cap N_2 \cap \ldots \cap N_n)$ by Corollary 6.7. Thus we see that $D \neq D'$ if and only if $h(D \cap D') = 0$ if and only if $(J_1 \cap J_1') \cap N_2 \cap \ldots \cap N_n = 0$ if and only if $J_1 \cap J_1' = N_1$. It follows that K has only one simple divisible submodule corresponding to P_1 if and only if N_1 is an irreducible ideal.

(3) <=> (2). With the notation of the preceding paragraph, let $\mathcal{S} = H - (P_1 \cup \ldots \cup P_n)$. Then $H_{\mathcal{S}}$ is the full ring of quotients of H, and hence $H_{\mathcal{S}} \cong H \otimes_R Q$. It is easy to see that $H_{\mathcal{S}} \cong H_{P_1} \oplus \ldots \oplus H_{P_n}$. Thus $H_{\mathcal{S}}$ is self-injective if and only if every H_{P_i} is self-injective. Furthermore, N_i is an irreducible ideal of H if and only if 0 is an irreducible ideal of H_{P_i}. Thus, since H_{P_i} is an Artinian local ring, it is sufficient to prove that an Artinian local ring S is self-injective if and only if 0 is an irreducible ideal of S.

If S is self-injective, then S is an indecomposable injective S-module. Because the annihilator of 1 in S is the 0 ideal, it follows from Theorem 4.5 that 0 is an irreducible ideal of S. Conversely, suppose that 0 is an irreducible ideal of S. If P is the maximal ideal of S, then 0 is an irreducible P-primary ideal of S, and hence $E(S) = E(S/P)$ by Theorem 4.5. Since $\mathrm{Hom}_S(S/P, E(S/P)) \cong S/P$, it follows easily by induction on length that if A is any S-module of finite length, then $\mathcal{L}(A) = \mathcal{L}(\mathrm{Hom}_S(A, E(S/P)))$. Since S has finite length over S, we see that $\mathcal{L}(S) = \mathcal{L}(\mathrm{Hom}_S(S, E(S/P))) = \mathcal{L}(E(S/P))$. Because

$S \subset E(S) = E(S/P)$, we see that $S = E(S/P)$, and hence S is self-injective.

Remarks. (1) The equivalence of several of the statements in Theorems 15.5 and 15.7 has been proved by a different method in [7].

(2) We have seen in Theorem 13.1 that R is a canonical ideal for R if and only if R is a Gorenstein ring; that is, if and only if K is isomorphic to E. Theorem 15.7 shows that in general R has a canonical ideal if and only if K is equivalent to E. In the proof of Theorem 13.1 we established that R is a Gorenstein ring if and only if there is a monomorphism of K into E, while Theorem 15.7 shows that R has a canonical ideal if and only if there is an epimorphism of K onto E.

(3) It follows from Theorem 15.7 that if S is an R-submodule of Q such that $R \subset S$ and $Q/S \cong E$, then S is isomorphic to a regular ideal of R.

(4) Theorem 10.3 states that if the integral closure of R in Q is a finitely generated R-module, then K is equivalent to E. Hence by Theorem 15.7 we see that if R is analytically unramified, then R has a canonical ideal.

Theorem 15.8. *Any two canonical ideals of R (if any exist) are isomorphic.*

Proof. Suppose that I and J are canonical ideals of R. Since J is a regular ideal of R, there is a regular element b in J. Now $bI \subset J$, and bI is a canonical ideal isomorphic to I. Hence without loss of generality we can assume that $I \subset J$. Let $B = J/I$; then because $Q/I \cong E$ and $Q/J \cong E$ we have an exact sequence:

$$0 \to B \to E \to E \to 0$$

Let $B^{\#} = \mathrm{Hom}_R(B,E)$; then since $\mathrm{Hom}_R(E,E) \cong H$, we can derive an exact sequence:

$$0 \to H \xrightarrow{\alpha} H \to B^{\#} \to 0.$$

By Theorem 2.7, α is an H-homomorphism, and thus α is multiplication by a regular element g of H. Hence we have $B^{\#} \cong H/Hg$.

Let $L = Hg \cap R$; by Theorem 2.8 we have $H \otimes_R L \cong HL = Hg$ and $H/Hg \cong R/L$. Since $H \otimes_R L$ is a projective ideal of H, it follows from Corollary 2.6 (2) that L is a projective ideal of R. Projective ideals in a local ring are principal, and hence there exists a regular element $a \in R$ such that $L = Ra$. Thus $H/Hg \cong R/Ra$.

Since $B^{\#} \cong H/Hg \cong R/Ra$, we have $aB^{\#} = 0$. By Theorem 4.6 we have $B \cong B^{\#\#}$, and hence it follows that $aB = 0$. Thus we see that $aJ \subset I$. Now B is an R-module of finite length. As we observed in the proof of Theorem 13.1, this implies that $\mathcal{L}(B) = \mathcal{L}(B^{\#})$. Therefore, $\mathcal{L}(B) = \mathcal{L}(R/Ra)$. By Theorem 12.3, $\mathcal{L}(R/Ra) = \mathcal{L}(J/Ja)$, and hence $\mathcal{L}(J/I) = \mathcal{L}(B) = \mathcal{L}(J/Ja)$. Because $Ja \subset I$ it follows that $Ja = I$. Therefore, I and J are isomorphic ideals of R.

REFERENCES

(1) H. Bass, "On the ubiquity of Gorenstein rings," Math. Zeitschr., 82 (1963), 8-28.

(2) N. Bourbaki, "Elements de Mathématique, Algèbre Commutative", Fascicule XXVII, No. 1290, Hermann, Paris (1961).

(3) H. Cartan and S. Eilenberg, "Homological Algebra", Princeton University Press, Princeton, N. J. (1956).

(4) J. Dieudonné and A. Grothendieck, "Eléments de Géométrie Algébrique I", Springer-Verlag, Berlin,Heidelberg, New York (1971)

(5) D. Ferrand and M. Raynaud, "Fibres Formelles d'un Local Noethérien", Ann. Scient. Ec. Norm. Sup. 4e Serie, t.3, (1970), 295-311.

(6) R. Hamsher, "On the structure of a one dimensional quotient field", J. of Algebra, 19 (1971), 416-425.

(7) J. Herzog and E. Kunz, "Der Kanonische Modul eines Cohen-Macaulay-Rings", Lecture Notes in Mathematics No. 238, Springer-Verlag, Berlin, Heidelberg, New York (1971).

(8) J. Lipman, "Stable ideals and Arf rings", Am. J. of Math., Vol. XCIII, No. 3, (1971), 649-685.

(9) E. Matlis, "Injective modules over Noetherian rings", Pacific J. Math., 8 (1958), 511-528.

(10) E. Matlis, "Divisible modules", Proc. Amer. Math. Soc., 11 (1960) 385-391.

(11) E. Matlis, "Some properties of a Noetherian domain of dimension 1", Canadian J. Math., 13 (1961), 569-586.

(12) E. Matlis, "Cotorsion modules", Memoirs Amer. Math. Soc., No. 49, (1964).

(13) E. Matlis, "Decomposable modules", Trans. Amer. Math. Soc., 125 (1966), 147-179.

(14) E. Matlis, "Reflexive Domains", J. Algebra, 8 (1968), 1-33.

(15) E. Matlis, "The multiplicity and reduction number of a one-dimensional local ring", Proc. London Math. Soc., (to appear).

(16) E. Matlis, "The theory of Q-rings", (to appear).

(17) M. Nagata, "Local rings", Interscience Publishers, New York, N. Y., (1962).

(18) D. G. Northcott, "Ideal Theory", Cambridge Tracts, No. 42, Cambridge University Press, London (1953).

(19) D. G. Northcott, "General theory of one-dimensional local rings" Proc. Glasgow Math. Assoc., 2 (1956), 159-169.

(20) D. G. Northcott, "On the notion of a first neighborhood ring", Proc. Camb. Phil. Soc., 53 (1957), 43-56.

(21) D. G. Northcott, "The theory of one-dimensional rings", Proc. London Math. Soc., 8 (1958), 388-415.

(22) D. G. Northcott, "The reduction number of a one-dimensional local ring", Mathematica, 6 (1959), 87-90.

(23) D. G. Northcott, "An algebraic relation connected with the theory of curves", J. London Math. Soc., 34 (1959), 195-204.

(24) O. Zariski and P. Samuel, "Commutative Algebra", Vol. II, Van Nostrand, Inc., Princeton, N. J. (1960).

INDEX

Please turn over

Vol. 212: B. Scarpellini, Proof Theory and Intuitionistic Systems. VII, 291 pages. 1971. DM 24,-

Vol. 213: H. Hogbe-Nlend, Théorie des Bornologies et Applications. V, 168 pages. 1971. DM 18,-

Vol. 214: M. Smorodinsky, Ergodic Theory, Entropy. V, 64 pages. 1971. DM 16,-

Vol. 215: P. Antonelli, D. Burghelea and P. J. Kahn, The Concordance-Homotopy Groups of Geometric Automorphism Groups. X, 140 pages. 1971. DM 16,-

Vol. 216: H. Maaß, Siegel's Modular Forms and Dirichlet Series. VII, 328 pages. 1971. DM 20,-

Vol. 217: T. J. Jech, Lectures in Set Theory with Particular Emphasis on the Method of Forcing. V, 137 pages. 1971. DM 16,-

Vol. 218: C. P. Schnorr, Zufälligkeit und Wahrscheinlichkeit. IV, 212 Seiten 1971. DM 20,-

Vol. 219: N. L. Alling and N. Greenleaf, Foundations of the Theory of Klein Surfaces. IX, 117 pages. 1971. DM 16,-

Vol. 220: W. A. Coppel, Disconjugacy. V, 148 pages. 1971. DM 16,-

Vol. 221: P. Gabriel und F. Ulmer, Lokal präsentierbare Kategorien. V, 200 Seiten. 1971. DM 18,-

Vol. 222: C. Meghea, Compactification des Espaces Harmoniques. III, 108 pages. 1971. DM 16,-

Vol. 223: U. Felgner, Models of ZF-Set Theory. VI, 173 pages. 1971. DM 16,-

Vol. 224: Revètements Etales et Groupe Fondamental. (SGA 1). Dirigé par A. Grothendieck XXII, 447 pages. 1971. DM 30,-

Vol. 225: Théorie des Intersections et Théorème de Riemann-Roch. (SGA 6). Dirigé par P. Berthelot, A. Grothendieck et L. Illusie. XII, 700 pages. 1971. DM 40,-

Vol. 226: Seminar on Potential Theory, II. Edited by H. Bauer. IV, 170 pages. 1971. DM 18,-

Vol. 227: H. L. Montgomery, Topics in Multiplicative Number Theory. IX, 178 pages. 1971. DM 18,-

Vol. 228: Conference on Applications of Numerical Analysis. Edited by J. Ll. Morris. X, 358 pages. 1971. DM 26,-

Vol. 229: J. Väisälä, Lectures on n-Dimensional Quasiconformal Mappings. XIV, 144 pages. 1971. DM 16,-

Vol. 230: L. Waelbroeck, Topological Vector Spaces and Algebras. VII, 158 pages. 1971. DM 16,-

Vol. 231: H. Reiter, L^1-Algebras and Segal Algebras. XI, 113 pages. 1971. DM 16,-

Vol. 232: T. H. Ganelius, Tauberian Remainder Theorems. VI, 75 pages. 1971. DM 16,-

Vol. 233: C. P. Tsokos and W. J. Padgett. Random Integral Equations with Applications to Stochastic Systems. VII, 174 pages. 1971. DM 18,-

Vol. 234: A. Andreotti and W. Stoll. Analytic and Algebraic Dependence of Meromorphic Functions. III, 390 pages. 1971. DM 26,-

Vol. 235: Global Differentiable Dynamics. Edited by O. Hájek, A. J. Lohwater, and R. McCann. X, 140 pages. 1971. DM 16,-

Vol. 236: M. Barr, P. A. Grillet, and D. H. van Osdol. Exact Categories and Categories of Sheaves. VII, 239 pages. 1971, DM 20,-

Vol. 237: B. Stenström. Rings and Modules of Quotients. VII, 136 pages. 1971. DM 16,-

Vol. 238: Der kanonische Modul eines Cohen-Macaulay-Rings. Herausgegeben von Jürgen Herzog und Ernst Kunz. VI, 103 Seiten. 1971. DM 16,-

Vol. 239: L. Illusie, Complexe Cotangent et Déformations I. XV, 355 pages. 1971. DM 26,-

Vol. 240: A. Kerber, Representations of Permutation Groups I. VII, 192 pages. 1971. DM 18,-

Vol. 241: S. Kaneyuki, Homogeneous Bounded Domains and Siegel Domains. V, 89 pages. 1971. DM 16,-

Vol. 242: R. R. Coifman et G. Weiss, Analyse Harmonique Non-Commutative sur Certains Espaces. V, 160 pages. 1971. DM 16,-

Vol. 243: Japan-United States Seminar on Ordinary Differential and Functional Equations. Edited by M. Urabe. VIII, 332 pages. 1971. DM 26,-

Vol. 244: Séminaire Bourbaki - vol. 1970/71. Exposés 382-399. IV, 356 pages. 1971. DM 26,-

Vol. 245: D. E. Cohen, Groups of Cohomological Dimension One. V, 99 pages. 1972. DM 16,-

Vol. 246: Lectures on Rings and Modules. Tulane University Rin and Operator Theory Year, 1970-1971. Volume I. X, 661 pages. 197 DM 40,-

Vol. 247: Lectures on Operator Algebras. Tulane University Ring an Operator Theory Year, 1970-1971. Volume II. XI, 786 pages. 197 DM 40,-

Vol. 248: Lectures on the Applications of Sheaves to Ring Theor Tulane University Ring and Operator Theory Year, 1970-1971. Vo ume III. VIII, 315 pages. 1971. DM 26,-

Vol. 249: Symposium on Algebraic Topology. Edited by P. J. Hilto VII, 111 pages. 1971. DM 16,-

Vol. 250: B. Jónsson, Topics in Universal Algebra. VI, 220 page 1972. DM 20,-

Vol. 251: The Theory of Arithmetic Functions. Edited by A. A. Gio and D. L. Goldsmith VI, 287 pages. 1972. DM 24,-

Vol. 252: D. A. Stone, Stratified Polyhedra. IX, 193 pages. 197 DM 18,-

Vol. 253: V. Komkov, Optimal Control Theory for the Damping Vibrations of Simple Elastic Systems. V, 240 pages. 1972. DM 20,

Vol. 254: C. U. Jensen, Les Foncteurs Dérivés de $\varprojlim$ et leurs A plications en Théorie des Modules. V, 103 pages. 1972. DM 16,

Vol. 255: Conference in Mathematical Logic - London '70. Edited b W. Hodges. VIII, 351 pages. 1972. DM 26,-

Vol. 256: C. A. Berenstein and M. A. Dostal, Analytically Unifor Spaces and their Applications to Convolution Equations. VII, 13 pages. 1972. DM 16,-

Vol. 257: R. B. Holmes, A Course on Optimization and Best A proximation. VIII, 233 pages. 1972. DM 20,-

Vol. 258: Séminaire de Probabilités VI. Edited by P. A. Meyer. V 253 pages. 1972. DM 22,-

Vol. 259: N. Moulis, Structures de Fredholm sur les Variétés H bertiennes. V, 123 pages. 1972. DM 16,-

Vol. 260: R. Godement and H. Jacquet, Zeta Functions of Simp Algebras. IX, 188 pages. 1972. DM 18,-

Vol. 261: A. Guichardet, Symmetric Hilbert Spaces and Related T pics. V, 197 pages. 1972. DM 18,-

Vol. 262: H. G. Zimmer, Computational Problems, Methods, ar Results in Algebraic Number Theory. V, 103 pages. 1972. DM 16

Vol. 263: T. Parthasarathy, Selection Theorems and their Application VII, 101 pages. 1972. DM 16,-

Vol. 264: W. Messing, The Crystals Associated to Barsotti-Ta Groups: with Applications to Abelian Schemes. III, 190 pages. 197 DM 18,-

Vol. 265: N. Saavedra Rivano, Catégories Tannakiennes. II, 4 pages. 1972. DM 26,-

Vol. 266: Conference on Harmonic Analysis. Edited by D. Guli and R. L. Lipsman. VI, 323 pages. 1972. DM 24,-

Vol. 267: Numerische Lösung nichtlinearer partieller Differential- u Integro-Differentialgleichungen. Herausgegeben von R. Ansorge u W. Törnig, VI, 339 Seiten. 1972. DM 26,-

Vol. 268: C. G. Simader, On Dirichlet's Boundary Value Problem. IV, pages. 1972. DM 20,-

Vol. 269: Théorie des Topos et Cohomologie Etale des Schém (SGA 4). Dirigé par M. Artin, A. Grothendieck et J. L. Verdier. X 525 pages. 1972. DM 50,-

Vol. 270: Théorie des Topos et Cohomologie Etle des Schém Tome 2. (SGA 4). Dirige par M. Artin, A. Grothendieck et J. L. Verdi V, 418 pages. 1972. DM 50,-

Vol. 271: J. P. May, The Geometry of Iterated Loop Spaces. IX, 1 pages. 1972. DM 18,-

Vol. 272: K. R. Parthasarathy and K. Schmidt, Positive Definite K nels, Continuous Tensor Products, and Central Limit Theorems Probability Theory. VI, 107 pages. 1972. DM 16,-

Vol. 273: U. Seip, Kompakt erzeugte Vektorräume und Analysis. 119 Seiten. 1972. DM 16,-

Vol. 274: Toposes, Algebraic Geometry and Logic. Edited by. F. Lawvere. VI, 189 pages. 1972. DM 18,-

Vol. 275: Séminaire Pierre Lelong (Analyse) Année 1970-1971. 181 pages. 1972. DM 18,-

Vol. 276: A. Borel, Représentations de Groupes Localement Co pacts. V, 98 pages. 1972. DM 16,-

Vol. 277: Séminaire Banach. Edité par C. Houzel. VII, 229 pa 1972. DM 20,-